BIOCHEMISTRY RESEARCH TRENDS

HYALURONAN

BIOLOGICAL AND MEDICAL IMPLICATIONS

BIOCHEMISTRY RESEARCH TRENDS

Additional books in this series can be found on Nova's website
under the Series tab.

Additional e-books in this series can be found on Nova's website
under the e-book tab.

BIOCHEMISTRY RESEARCH TRENDS

HYALURONAN

BIOLOGICAL AND MEDICAL IMPLICATIONS

VITOR H. POMIN

EDITOR

nova publishers

New York

Library of Congress Cataloging-in-Publication Data

ISBN: 978-1-63117-808-5

Published by Nova Science Publishers, Inc. † New York

CONTENTS

PREFACE

Hyaluronan or hyaluronic acid (HA), the only non-sulfated glycosaminoglycan (GAG) of the body, is structurally composed of disaccharide repeating units of alternating 3-linked N-acetyl β-D-glucosamine and 4-linked β-D-glucuronic acid. Despite the lack of sulfation and the lowest structural heterogeneity among all GAGs, HA still retains efficient capability in interacting with many molecules, especially functional proteins like growth factors, cytokines, and adhesive proteins involved in angiogenesis, tumor proliferation, migration, and invasion. However, the main biological role of HA is to assemble the extracellular matrices of certain soft tissues, such as vitreous humor, synovial fluid, umbilical cord, and the intercellular space of the epidermis. HA is the major structure-forming component in the globe of the eye and cornea. For this reason, this GAG can be used to compose ophthalmic solutions. The most famous physicochemical property of HA is its capacity to make gels in solution. This property can be pharmaceutically and industrially useful as either a vehicle to make specific media or biologically active sera. This book, heavily based on recent scientific results from research groups of several countries, will cover many of these aspects of HA, the related enzymes and metabolism in health and disease. The seven chapters are basically centered in 1) the covalent modifications of HA by serum protein and the biological consequences; 2) the aspects of HA as a function of age; 3) carcinogenesis as a function of HA metabolism; 4) therapeutic applications of HA-related enzymes; 5) characterization and analyses of HA-based ophthalmic solutions; 6) the role of HA in reproductive medicine; and 7) the role of HA in bone medicine. The contributing authors are renowned scientists in the HA field and this collaborates to the high-quality of this publication.

Chapter 1 - During inflammatory responses and cancer malignance, hyaluronan becomes covalently modified by serum-derived hyaluronan-associated proteins (SHAP), which are derived from the heavy chains of inter-alpha-trypsin inhibitor (ITI). The physiological significance of SHAP modification of hyaluronan was explored using knockout mice deficient in SHAP-hyaluronan complex formation. In ovalbumin-induce airway hyperresponsiveness (AHR) model, the SHAP-deficient mice exhibited elevated AHR. The levels of sTNFR1 and IL12p40 in the airway secretion were greatly reduced in SHAP-deficient mice. In addition, the proliferation of epithelial cell and the repair of airway epithelium were significantly impaired in SHAP-deficient mice. These findings suggest that the SHAP modification of hyaluronan is important to the protection of airway epithelium and remission of tracheal inflammation. In contrast, the SHAP-deficient mice were resistant to bacterial lipopolysaccharide (LPS)/D-galactosamine-induced acute liver injury, exhibiting significant decrease in the serum alanine aminotransferase level, the number of apoptotic hepatocytes and infiltrated neutrophils, and the mortality rate. LPS stimulation neither upregulate the avidity of neutrophil CD44 to HA nor increase the amount of hyaluronan on the surface of liver sinusoidal endothelium. But it stimulated the deposition of SHAP on sinusoidal HA. Cell adhesion experiment showed that the presence of SHAP dramatically potentiates the adhesion of CD44-bearing cells to hyaluronan. Consistently, LPS-induced neutrophil adhesion to liver sinusoids was reduced by 50% in SHAP-deficient mice. Therefore, SHAP-HA-CD44 interaction underlies LPS-induced liver injury. Taken together, these findings show clearly that the SHAP modification of hyaluronan is crucial to hyaluronan function in inflammatory responses. However, the effect could be either proinflammatory or anti-inflammatory, depending on the tissues and cells involved.

Chapter 2 - The effect of aging on Hyaluronan (HA) has been studied primarily in the context of skin and cutaneous wound repair in animal models ranging from mice to human. Fetal skin contains high levels of higher molecular weight (HMW) HA, which is well described as inducing healing without fibrosis and scar formation. Adult skin is thought to retain HA content, but have alterations in HA synthesis. There is also evidence that aging alters cleavage of native HMW HA into lower MW forms. For example, author have found that age-related impairments in cutaneous wound repair are correlated with a deficit in the production of LMW-HA in aged mouse dermis. More recently, author have demonstrated that healing of dermal wounds in aged mice is significantly improved by treatment with HA of lower MW size. The influence of aging on HA content and MW in other tissues remains incompletely characterized. Aging might affect the extraction of HA, thereby altering the accurate measurement of HA content and size. Data in other organs such as the brain, prostate, and joints have evaluated HA content and size in the context of diseases that are more common with aging, rather than age itself. These studies also suggest that with aging, most tissues modify HA content, size and influence on cellular processes relevant to injury, repair, and the development of pathology. This chapter will provide an overview of age-related changes in HA in the basal state and during injury in organs such as skin, brain, prostate and skeletal joints.

Chapter 3 - Most of the cells in multicellular organisms are in contact with intricate meshwork of interacting, extracellular macromolecules that constitute extracellular matrix (ECM). The ECM plays an important role not only in tissue architecture but also in regulation of gene expression and cell behavior. Many of the ECM effects are mediated by dynamic interaction between ECM macromolecules and their receptors. Among the ECM components, Hyaluronic Acid (HA) is one of the fascinating molecule. HA is a high molecular weight, uniformly repetitive, linear glycosaminoglycan (GAG) composed of disaccharides of glucuronic acid and N-acetylglucosamine. HA is huge-not only in molecular weight, but also in the space it occupies in solution, and its versatile biological function is completely dependent on its size. Earlier studies showed that cell divide and migrate within ECM rich in HA. The dramatic modulation of HA concentration and organization accompany cellular changes that take place during tissue and organ differentiation. In recent years, ample of studies have revealed the critical role of HA in carcinogenesis in many types of cancer, and it can be an independent prognostic indicator. During tumor development, cancerous cells tend to synthesize more HA and has been shown to accumulate in the intracellular compartments and pericellular matrices. The accumulated HA in the tumor is known to triggers signaling pathways that are favorable for tumor growth. These effects are mediated through the interaction between HA and its binding proteins (HABPs, Hyaluronic Acid Binding Proteins). Further, catabolism of HA to HA oligosaccharides by hyaluronidase enzyme leads to angiogenesis and new array of cell signaling events. Therefore, targeting aberrant HA synthesis and accumulation could potentially prevent tumor growth. In this chapter, author will discuss the biological synthesis of hyaluronic acid during normal developmental process and carcinogenesis. Further author will discuss the catabolism of HA to HA oligosaccharides. Finally, author will discuss the possible ways to prevent aberrant HA synthesis and its accumulation during carcinogenesis.

Chapter 4 - Hyaluronidases are enzymes able to degrade hyaluronic acid. They are classified in three groups: mammalian, leeches and microbial hyaluronidases. Their role in vivo is essential for the homeostasis and metabolism of the extracellular matrix, thus regulating cell growth. Their action could be measured by chemical, physicochemical and biological methods. The first employ reductimetric reactions on the products of hyaluronidases. Physicochemical procedures measure the changes of physical features of the substrate, like viscosity. Biological methods evaluate the spreading effect of the enzyme in the animal model. Several applications have been described in literature, mainly a spreading factor. The most important application is hypodermoclysis, the capacity of increasing absorption and dispersion of an injected drug. Hyaluronidases are employed in radiography for increasing the spread of radiopaque substances, in gynecology along with ergometrine to prevent post-partum hemorrhage, in obstretric blocks of the pudendal and ileoinguinal nerve. In ophtalmology hyaluronidases are instilled with local anesthetics for retrobulbar, peribulbar, sub-Tenon's, and van Lint blocks. In fact the injection of hyaluronidase increases in intraocular pressure (IOP), less distortion of the surgical site, decreased incidence of postoperative strabismus, and potential for limiting local anesthetic myotoxicity. Hyaluronidase could also be employed for post-operative edema reduction. In this perspective hyaluronidases are employed in transplant surgery as long as they reduce interstitial edema and the risk of rejection. In chemotherapy hyaluronidases prevent the risk of local injury after drug extravasation. In pain therapy the enzymes are employed to increase the spread of local anesthetics in selective blocks, while in cardiology they could reduce the size of myocardial infarcts after coronary occlusion. Moreover hyaluronidases can be successfully employed to treat complications by inappropriate subcutaneous injection of hyaluronic acid for aesthetic purposes. In fact an excessive inoculation of hyaluronic acid filler mayhesitate in visible overecorrection, nodules, bumps or ischemic complications by compression of the dermal plexus. Given the importance of these enzymes for homeostasis and their applications in Medicine, a proper knowledge is essential for the everyday practice.

Chapter 5 - Hyaluronan (HA), also known as hyaluronic acid or sodium hyaluronate, has drawn considerable interest and attention from ophthalmic surgical and eye care industries due to its biocompatibility, lubricity, and viscoelastic properties. A more complete understanding of HA biopolymers has therefore become increasingly critical since thorough characterization of raw materials and finished ophthalmic products helps promote product quality and process control. Often such detailed information requires the use of a combination of analytical techniques. In this chapter, author compare size exclusion chromatography (SEC) with on-line multi-angle light scattering (SEC-MALS) and SEC with triple detection (SEC-TD) experiments for HA analysis. Three lots of commercially available eye drop grade HA raw materials were characterized by SEC-MALS and SEC-TD. The absolute molecular weight averages, molecular weight distribution, radius of gyration, intrinsic viscosity, and solution conformation of the HA lots were determined and compared by the two techniques.

In addition, the molecular weights and concentrations of HA in eleven marketed ophthalmic products were evaluated by hyaluronidase digestion experiments and by SEC-TD. The weight-average molecular weights (M_w) of the products tested ranged from 155,000 to 1,400,000 Daltons and the concentration of HA ranged from 0.003% to 0.15%. The work presented in this chapter is from the first publication on characterizing HA molecular weights and concentrations in various marketed HA containing ophthalmic products. Research has indicated that there is a correlation between the molecular weight of hyaluronan and its biocompatibility/biological functions, with high molecular weight HA exhibiting many biological and physiological benefits. Future investigation of the effect of low molecular weight HA on eye is required.

Chapter 6 - This chapter presents the usefulness of hyaluronan in reproductive medicine. Hyaluronan is a major constituent of the extracellular matrix of the cumulus cells in the cumulus-oocyte complex and may play a critical role in the selection of functionally competent spermatozoa during *in vivo* or *in vitro* fertilization. Hyaluronan can serve as the selective marker during intracytosplasmic sperm injection (ICSI) to select and inject the most optimal spermatozoon. The technique of ICSI requires the immobilization of the spermatozoa. Polyvinylpyrrolidone (PVP) routinely used during ICSI, facilitates handling of spermatozoa. PVP is an artificial polymer, which has been regarded as chemically inert, although adverse effects as a result of its use have been reported. Viscous solution of hyaluronan can be used as an alternative to toxic PVP. Moreover, hyaluronan is also a major glucosaminoglycan in the uterine fluid. It has been shown to increase cell-cell adhesion and may improve embryo apposition and attachment. Some studies suggest that the use of hyaluronan containing embryo transfer medium can improve the implantation.

Chapter 7 - Bone serves as a structural framework and provides protection to various organs, yet also contributes substantially to hemopoiesis and mineral homeostasis. Osseous tissue is richly supplied with blood vessels and nerves, which are embedded in extracellular matrix composed of collagen fibers and crystallized calcium-phosphate salts known as hydroxyapatite. After a bone fracture, typically blood leaks from torn ends of vessels and forms a clotted mass known as the fracture hematoma. During healing progression, fibroblasts, chondroblasts, and osteogenic cells invade the fracture site and start forming a cartilage-and hyaluronan-rich fibrous callus that is replaced first by spongy bone trabeculae and subsequently further remodeled to compact bone. Depending on the site and size of the bone defect, but also on the health status and age of the patient, complete healing can take up to several months. In this regard, treatment with β-tricalcium phosphate-based bone graft substitutes has been reported to support the overall regeneration progress. Further along these lines, hyaluronan-coated biocompatible bone replacement material is providing a more favorable environment for invading cells or was shown to serve as a delivery vehicle for autologous cells. In this review the osteoconductive role of hyaluronan and its medical application in bone graft substitutes in the context of fracture healing, bone fragment fixation surgery and implant osseointegration will be discussed.

In: Hyaluronan
Editor: Vitor H. Pomin

ISBN: 978-1-63117-808-5
© 2014 Nova Science Publishers, Inc.

Chapter 1

COVALENT MODIFICATION OF HYALURONAN BY SERUM PROTEIN AND ITS BIOLOGICAL CONSEQUENCE

Lisheng Zhuo and **Koji Kimata***
Advanced Medical Research Center,
Aichi Medical University, Nagakute, Aichi, Japan

ABSTRACT

During inflammatory responses and cancer malignance, hyaluronan (HA) becomes covalently modified by serum-derived hyaluronan-associated proteins (SHAP), which are derived from the heavy chains of inter-alpha-trypsin inhibitor (ITI). The physiological significance of SHAP modification of HA was explored using knockout mice deficient in SHAP-HA complex formation. In ovalbumin-induce airway hyperresponsiveness (AHR) model, the SHAP-deficient mice exhibited elevated AHR. The levels of sTNFR1 and IL12p40 in the airway secretion were greatly reduced in SHAP-deficient mice. In addition, the proliferation of epithelial cell and the repair of airway epithelium were significantly impaired in SHAP-deficient mice. These findings suggest that the SHAP modification of HA is important to the protection of airway epithelium and remission of tracheal inflammation. In contrast, the SHAP-deficient mice were resistant to bacterial lipopolysaccharide

---

* Email address: zhuols@gmail.com.

(LPS)/D-galactosamine-induced acute liver injury, exhibiting significant decrease in the serum alanine aminotransferase level, the number of apoptotic hepatocytes and infiltrated neutrophils, and the mortality rate. LPS stimulation neither upregulate the avidity of neutrophil CD44 to HA nor increase the amount of HA on the surface of liver sinusoidal endothelium. But it stimulated the deposition of SHAP on sinusoidal HA. Cell adhesion experiment showed that the presence of SHAP dramatically potentiates the adhesion of CD44-bearing cells to HA. Consistently, LPS-induced neutrophil adhesion to liver sinusoids was reduced by 50% in SHAP-deficient mice. Therefore, SHAP-HA-CD44 interaction underlies LPS-induced liver injury. Taken together, these findings show clearly that the SHAP modification of HA is crucial to HA function in inflammatory responses. However, the effect could be either proinflammatory or anti-inflammatory, depending on the tissues and cells involved.

INTRODUCTION

Hyaluronan (HA) occurs mainly as a component of the extracellular matrix (ECM) and play important roles in supporting active cell propagation and migration. The expression of HA is high in fetal tissues that undergo active development, but decreases to a low level during adulthood in most tissues except a few loose connective tissues, for example skin, cartilage and vitreous body, where HA is a constitutive component [1]. However, when wounded, the tissues restart to the produce HA actively to assist the tissue repair. The relationship between HA upregulation and inflammation has been well established in various tissues under various conditions [2-3]. The serum level of HA is widely used as a clinical biomarker for the diagnosis of liver cirrhosis and for the monitoring of activity of rheumatoid arthritis.

In contrast to other glycosaminoglycans, HA is neither attached to a core protein nor sulfated. In spite of its structural simplicity, HA exhibits versatile function via interacting with over twenty kinds of proteins ranging from extracellular matrix component to intracellular molecules. Among them, serum-derived HA-associated protein (SHAP) is quite unique in forming a covalent complex with HA [4].

SHAP is first discovered from the HA preparation isolated from the extracellular matrix of cultured mouse dermal fibroblast [5]. It is soon identified as the heavy chains of inter-alpha-trypsin inhibitor (ITI) family proteins, which are included in the serum supplemented to the culture medium [6]. ITI family proteins consist of a common light chain called bikunin, which is a typical proteoglycan molecule bearing a single chondroitin sulfate (CS) chain, and 1-2 heavy chains that link to the CS chain via an ester bond between their C-terminal aspartate and an internal GalNAc in CS (Figure 1A). The formation of the SHAP-HA complex is an enzyme-catalyzed transesterification reaction, where HA substitutes for CS to form an ester linkage to the heavy chains in a similar manner [4,7] (Figure 1B).

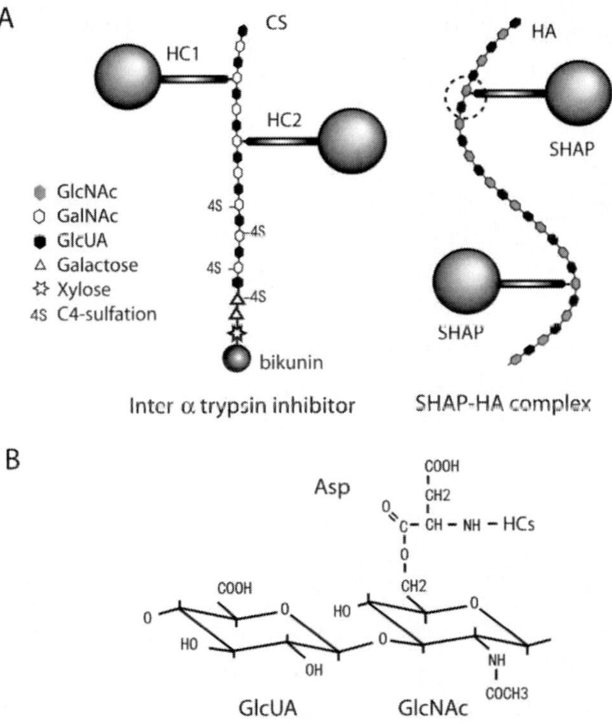

Figure 1. Schematic structures of inter-alpha trypsin inhibitor (ITI) and the SHAP-hyaluronan (HA) complex. A, ITI family proteins is composed of a common bikunin proteoglycan and 1-2 heavy chains (HCs) that link to the chondrotin sulfate (CS) chain of bikunin. The representative member, ITI is shown. The SHAP-HA complex is formed by a transesterification reaction where HA substitutes for CS to bind to HCs. B, The chemical structure of the SHAP-HA linkage region (dashed circle in A). The ester bond is formed between the carboxyl group of C-terminal aspartate of heavy chains and the C-6 hydroxyl group of an internal GlclNAc in HA. HC-CS linkage in ITI is via a similar ester bond.

ITI family proteins are secreted by hepatocytes and circulate at considerably high levels of 0.15-0.5 mg/ml in blood, where most of them bear the heavy chains. In contrast, the predominant member in urine is the heavy chain-free bikunin. It seems likely that the heavy chain moiety of ITI family proteins involves in biological processes by forming covalent complex with hyaluronan. Indeed, in vitro studies reveal that ITI stabilizes the HA-rich extracellular matrix of various cells lines and the isolated cumulus-oocyte complex [8]. The definitive evidence come from the study using SHAP-deficient knockout mice, which showed clearly that the formation of the SHAP-HA complex occurs within the expanding cumulus in preovulatory follicles and plays crtical roles in ovulation and fertilization [9].

Though the formation of the SHAP-HA complex can be achieved simply by incubating serum with HA, blood circulation is not the main stage for the complex formation because the half-life of HA in circulation is too short. During cumulus expansion, ITI effuses into the Graffian follicle to interact with the locally produced HA [10]. Similarly, in rheumatoid arthritis patients, the hyperplastic synovium actively produces HA, leading to the accumulation of a large amount of the SHAP-HA complex in the synovial fluid [11]. Since HA upregulation and plasma efflux are common events of inflammation, it is likely that HA produced in inflamed tissues is not free glycosaminogycans, but SHAP-associated. Such a molecular modification may underlie the reported functions of HA in inflammation. This chapter reviews the recent findings that support the notion.

AIRWAY HYPERRESPONSIVENESS

HA is an important component of the extracellular matrix in lung parenchyma. It is also a constituent of bronchial secretions released by the serous cells of submucosal glands of bronchus [12]. The alteration of HA in size, deposition and turnover has been extensively studied in various lung disorders. It was found to accumulate in the airway mucosa and bronchoalveolar lavage fluid (BALF) in asthma patients [13].

Asthma is a result of chronic airway inflammation that subsequently leads to airway hyperresponsiveness (AHR) with increased contractability of the surrounding smooth muscles and the bouts of airway narrowing and the classic symptoms of wheezing. Asthma is often accompanied by airway remodeling, including an invasion by inflammatory cells and myofibroblasts, an excessive proliferation of airway smooth muscle cells, an increase in the numbers of mucous glands, a thickening of the lamina reticularis, and a deposition of HA-rich extracellular matrix in bronchial mucosa [14-15]. The mucosal HA is thought to play important roles in mucus hydration, surface protection and epithelial repair, and thus the mucociliary clearance of foreign material out of the bronchial tree. In addition, HA may facilitate the migration of inflammatory cells and interact with the growth factors, proteases and protease inhibitors present in bronchial secretions [16].

The effect of SHAP on the function of HA in AHR pathogenesis was explored using the SHAP-deficient konckout mice [17]. AHR was induced with a classical method, i.e., the sensitization by injecting ovalbumin intraperitoneally and then the challenge with ovalbumin aerosol 30 min daily for 2 weeks. AHR was evaluated by measuring the enhanced pause (Penh) value, an indicator of the airway resistance. Under such conditions, the histological changes were comparable between the wild type mice and SHAP-deficient mice. However, the latter exhibited an airway resistance significantly higher than that of wild type mice when exposed to either the allergen ovalbumin or the contractile agent methacholine. Examination of HA in BALF confirmed the HA production and secretion were increased by AHR induction, and these HA molecules were SHAP-associated in the wild type mice but not in SHAP-deficient mice. These findings link up the HA modification by SHAP and AHR pathogenesis, suggesting a suppressive function of SHAP against the development of AHR.

The enhanced AHR in the absence of SHAP seems not related to eosinophil, the major inflammatory cells in asthmatic BALF, because its number and the serum level of allergen-specific IgE were not changed. In contrast, the numbers of neutrophil and macrophage in BALF were significantly increased in the absence of SHAP [17]. Examination of BALF cytokines showed that the levels of most cytokines (e.g. IL-4, IL-5, IL-6, IL-9, IL-10, IL12p70, IL-13, IFN-γ, TNF-α, and TGFβ1) were also comparable between the wild type mice and SHAP-deficient mice. However, significant difference was observed in the production of soluble TNF receptor 1 (sTNFR1) and IL12p40, which increased by 7 and 9 folds in the wild type mice, respectively, but only by 4.6 and 3 folds in SHAP-deficient mice [17].

TNF-α plays a central role in AHR development and airway remodeling [18]. TNF-α blockade effectively inhibit airway inflammation and AHR, and was proved to be effective in asthma treatment in a series of clinical trials [19]. sTNFR1, which blocks TNF receptor-mediated functions by competitively binding to TNF-α, is a well-recognized naturally occurring TNF-α blockade [20-21]. Accordingly, the decrease of sTNFR1 in the absence of SHAP results in a higher relative activity of TNF-α, which explains, at least partially, the enhanced AHR in SHAP-deficient mice. In contrast to BALF, the lung tissues of SHAP-deficient mice exhibited a stronger TNFR1 immunoreactivity than those of wild type mice, suggesting that SHAP does not affect the upregulation of TNFR1, but accelerates its shedding from cell surface [17].

IL12p40 is known as a component of the bioactive cytokines IL12 (IL12p70, a heterodimer of IL12p35 and IL12p40) and IL23 (a heterodimer of IL23p19 and IL12p40), which promote the differentiation of Th1 and Th17 lymphocytes, respectively. IL12p40 also exits in monomic and homodimeric forms, which inhibit IL12-mediated functions by competitively binding to the common receptor IL12Rβ1 [22-23]. Apparently, less IL12p40 results in a stronger IL12 signal, which may inhibit the progression of asthma by suppress Th2 differentiation. However, the truth is that SHAP-deficient mice with lower IL12p40 level develop more severe AHR. It noteworthy that the BALF levels of Th2 cytokines (IL-4, IL-5, IL-9, IL-10 and IL-13) and Th1 cytokines (TNF-α and IFN-γ) are not altered [17], suggesting a comparable Th2 polarization in SHAP-deficient mice. In addition, though the function of IL-12p40 as a negative feedback loop is widely appreciated, it also has other functions independent of IL-12 [24]. The exact mechanism need to be further explored.

Taken together, SHAP, by accelerating the shedding of TNF receptor and the formation of homodimeric form of IL12p40, acts as an inhibitory factor in the development of AHR.

AIRWAY EPITHELIAL REPAIR

Pulmonary epithelial injury is central to the pathogenesis of many lung diseases. Bronchial epithelial injury occurs chronically in asthmatic patients [25]. Epithelial repair encompasses cell proliferation, migration, and differentiation. All of these processes require cell interactions with the extracellular matrix. Immunohistochemical study revealed the accumulation of SHAP-HA complex in airway epithelium [17]. To explore more precisely the role of SHAP in bronchial epithelial repair, a simpler model was utilized where epithelial injury was induced by the intraperitoneal injection of a single dose of naphthalene. Naphthalene generates isolated bronchial epithelial injury through specific cytotoxicity on bronchial Clara cells, which resolves within 7–21 days [26]. Seven days after naphthalene injection, both large and small bronchia of wild type mice demonstrated an orderly epithelial cell pattern. In contrast, SHAP-deficient mice showed dysregulated recovery, with a loss of columnar epithelium, vacuolated cells, and epithelial squamous metaplasia [27]. In addition, BrdU incorporation experiment showed that the proliferation of airway epithelial cells was significantly decreased in SHAP-deficient mice

[27]. These observations show clearly that SHAP accelerates the bronchial epithelial repair.

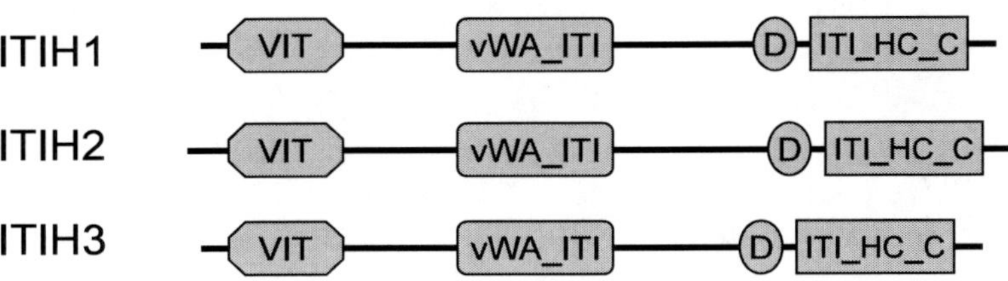

Figure 2 The domain structure of ITI HCs. HC1, HC2 and HC3 share high structural homology, consisting of vault protein inter-α-trypsin inhibitor (ITI) domain (VIT), von Willebrand factor type A ITI domain (vWA_ITI), and ITI heavy chain C-terminus domain (ITI_HC_C). The circled D indicates the aspartic acid residue that links to CS or HA. The coupling of HCs with bikunin involves the cleavage of the C-terminal one-third of the HC propeptide, whose fate is unknown, and the formation of an ester bond between the aspartic-acid residue and an internal N-acetyl galactosamine of chondroitin sulfate on bikunin.

The molecular mechanism was further explored. It was noted in immunohistochemical study that in airway epithelium the SHAP-HA complex colocalized largely with vitronectin, a well-known airway extracellular matrix component [27].

Vitronectin has been shown to protect bronchial epithelial cells from apoptosis, and promote cell adhesion, proliferation and migration via binding primarily to integrins [28-29]. Vitronectin deficiency impairs bronchial [28] and alveolar [30] epithelial repair. Some extracellular molecules, such as plasminogen activator inhibitor 1 (PAI-1), can modulate the cell-vitronectin interaction [31].

SHAP was thus examined for any effect on the cell-vitronectin interaction. Binding assay revealed a direct interaction between SHAP (possibly vWA domain) (Figure 2) and the RGD domain of vitronectin [27]. In addition, SHAP promoted cell adhesion onto vitronectin substratum and the vitronectin-dependent cell proliferation and migration in a cell culture system [27]. These observations are in well consistent with the in vivo findings in SHAP-deficient mice, and reveal that SHAP accelerates epithelial repair via enhancing vitronectin function (Figure 3).

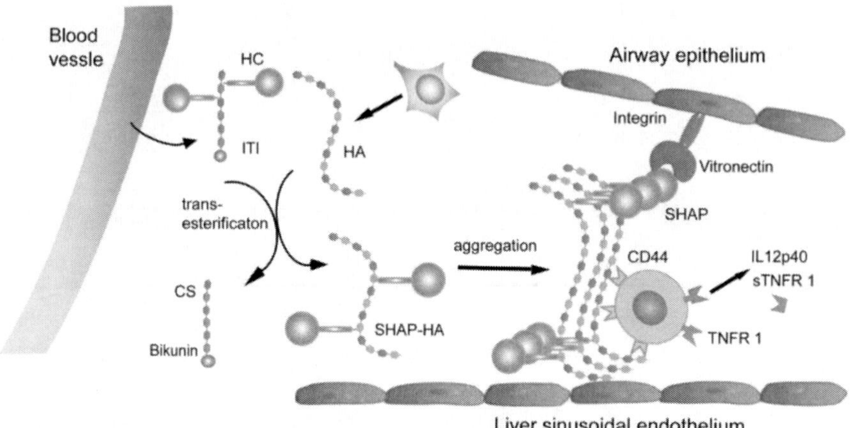

Figure 3. Schematic representation of the formation of SHAP-HA complex and its function. During inflammatory response, efflux of plasma proteins including ITI into inflamed tissues is induced. At the same time, the local production of HA is upregulated. The heavy chains of ITI are then transferred enzymatically onto HA to form the SHAP-HA complex. The reaction releases the bikunin moiety, which is then excreted into urine. SHAP induces the aggregation of HA, thus increases its avidity to CD44, resulting in the CD44-HA mediated adhesion of neutrophil to the liver sinusoids, which causes liver damage. On the other hand, SHAP formed in airway binds to vitronectin in the extracellular matrix of airway epithelium and enhances the epithelial protecting function of vitronectin, i. e., the prevention of apoptosis of bronchial epithelial cells and the promotion of cell adhesion, proliferation and migration. At the same time, SHAP accelerates, via yet unknown mechanisms, the shedding of TNFR1 and the production of IL12p40, which contribute to the suppression of airway allergic inflammation.

ACUTE LIVER INJURY AND NEUTROPHIL ADHESION

Adhesion molecules play important roles in hematopoesis, trafficking, recruitment, and the activation of leukocytes. The glycan-recognizing selectin family and the immunoglobulin superfamily molecules-recognizing integrin family are best known adhesion molecules, which mediate the primary adhesion and rolling of leukocytes on the endothelium and the subsequent firm adhesion, respectively.

HA and its cellular receptor CD44 emerged as novel adhesion molecules in 1990s [32]. CD44 was first cloned as a lymphocyte surface antigen recognized by lymphocyte homing-interfering monoclonal antibody. CD44 is a single polypeptide transmembrane protein encoded by a single gene, consisting of an N-terminal HA-binding domain, a stem region, a transmembrane domain and a cytoplasmic domain. The stem region shows a great variation as a result of alternative splicing of exons 6-15 (or exons V1-V10) of CD44 gene [33]. In addition, CD44 receives various chemical modifications, such as glycosylation (N-glycan, O-glycan, and glycosaminoglycan), sialylation, sulfation, phosphorylation, palmitoylation and dimerization/oligomerization [33-34]. The great structural diversity of CD44 is the prerequisite of the fine toning of CD44 function in a wide range of physiological and pathological events in various tissues.

In sharp contrast to CD44, the regulation of CD44-HA interaction on HA side has been neglected for long time because of its structural simplicity. Study with SHAP-deficient mice revealed that such a regulatory mechanism do exit and is important for the pathogenesis of acute liver injury.

Bacterial lipopolysaccharide (LPS) challenge induces pathophysiological reactions in various organs (including fever, hypotension, leukocyte infiltration and inflammation), resulting in a systemic syndrome resembling human septic shock in mice [35]. D-galactosamine (D-Gal) is a hepatocyte-specific toxin that ceases the biosysthesis of macromolecules (such as RNA, protein, glycoprotein and glycogen) and eventually causes the death of hepatocytes. The toxicity is a result of depletion of UTP, which is consumed in the reaction changing D-galactosamine to UDP-galactosamine derivatives. D-Gal treatment increases the susceptibility of mice to LPS by several thousand folds [35]. LPS toxicity in D-Gal-sensitized mice is defined to liver, mimicking the human fulminant hepatic failure [36-37].

TNF is the key molecule mediating LPS toxicity [38]. LPS stimulated the liver-resident macrophages (Kupffer cells) to produce TNF. Hepatocytes are normally resistant to TNF toxicity via intracellular protective factor(s) [39]. TNF at a proper concentration rather stimulates the proliferation of hepatocytes as exemplified by the mitogenic role of TNF in hepatic regeneration after partial hepatectomy. However, D-Gal treatment converts the TNF-induced cellular outcome from proliferative to apoptic [40]. The marked hepatic damage includes leukocyte recruitment, microvascular perfusion failure, caspase-3 activation, hepatocellular apoptosis, enzyme release and necrotic cell death.

SHAP-deficient mice exhibited a resistance against LPS/D-Gal-induced acute liver injury (our unpublished observation). The elevation of serum levels of TNF and alanine aminotransferase after LPS/D-Gal administration was significantly compromised. Immunohistochemical examination also revealed a decrease in the number of apoptotic hepatocytes and infiltrated neutrophil (by 50%) in the absence of SHAP. These resulted in a great improvement of surviving rate (from 0~20% in the wild type mice to 90% in the SHAP-deficient mice). These findings indicate a proinflammatory role of SHAP in GalN/LPS-induced acute liver injury.

Activation and recruitment of innate immune effector cells, namely neutrophils, within the microvasculature of liver is the key event of the immunopathogenesis of endotoxemia [41]. The recruited neutrophil causes vascular dysfunction, hepatocellular injury, and organ failure [42]. LPS-induced neutrophil adhesion to the postcapillary vessels in tissues such as muscle, mesentery, and skin is a classical paradigm involving selectin-mediated tethering and rolling, followed by integrin-mediated firm adhesion [43]. However, LPS-induced neutrophil recruitment to liver does not follow this paradigm. First, the neutrophils adhere not only to the postsinusoidal venules but also to the endothelium of capillaries (called sinusoids). Secondly, the adhesion to the sinusoids is a direct tethering and adhesion without a step of rolling, and is dependent on CD44-HA interaction but not selectin [44].

LPS treatment neither increases the avidity of CD44 on the surface of neutrophils to HA substratum, nor accelerates the deposition of HA on the sinusoids, but greatly enhances the accumulation of SHAP on sinusoidal HA [44]. This finding is in consistent with the observation that the number of adhered neutrophils in sinusoids is decreased by 50% in the absence of SHAP [45].

The effect of SHAP on CD44-HA interaction was further explored in detail in an in vitro cell adhesion assay system using SHAP-HA complex isolated from synovial fluid of rheumatoid arthritis patients [34]. The study showed that SHAP potentiates by folds the avidity of immobilized HA to CD44 on cell surface. Such an effect was abolished when the flow shear force exceeds 0.8 dyne/cm^2. The selectin-mediated adhesion is known to be able to endure stronger flow shear forces [46]. It is noteworthy that the blood flow in sinusoids is much slower than those in other vessels. It turns out that SHAP-HA/CD44 is chosen for neutrophil adhesion to sinusoids because the adhesion strength meets the blood flow condition there.

Biochemical analysis of the SHAP-HA complex found that SHAP promotes HA to aggregate into macromolecules [34]. Preferential binding of CD44-bearing cells to the aggregates was observed in cell adhesion assay as well as in the hyperplastic synovium from rheumatoid arthritis patients [34].

Collectively, these findings indicate that SHAP modifies the physicochemical and biological properties of HA, increasing its avidity to

CD44 by accelerating molecular aggregation. Such a mechanism underlies the LPS-induced neutrophil adhesion on liver sinusoids and the pathogenesis of acute liver injury (Figure 3).

CONCLUSION

This chapter reviews the recent study with SHAP-deficient mice. Abnormalities caused by SHAP deficiency were recognized in both the airway hyperresponsiveness and the acute liver injury models, providing convincing evidence for the notion that HA functions in the form of SHAP-HA complex in inflammatory responses. The role of SHAP-HA complex in inflammatory seems opposite in the two models, being proinflammatory in liver injury but anti-inflammatory in airway hyperresponsiveness. This is not a surprising because both HA and CD44 have shown such a discrepancy [17,34]. An inflammatory response is usually accomplished by cooperation of inflammatory factors and anti-inflammatory factors, which predominate the early and later phase of inflammation, respectively. The outcome of SHAP effect on inflammation differs based on what stage SHAP is involved, which is largely organ- and event- dependent.

The majority of HA studies have treated HA as a free glycosaminoglycan. From now on, the possibility of SHAP association should be awared of, especially when studying on a pathological role of HA. In addition, the medical application of chemically modified HA is now a very attractive and promising filed. SHAP represent an interesting modifier because it assigns novel physicochemical and biological properties to HA. Recently, the SHAP-HA complex is identified to be the effective component of amniotic membrane that is widely used as a surgical graft for ocular surface reconstruction and exerts clinically observable actions to promote epithelial wound healing and to suppress inflammation, scarring and angiogenesis [47].

REFERENCES

[1]   T. C. Laurent and J. R. Fraser, *FASEB J.* 6, 2397 (1992).
[2]   W. Y. Chen and G. Abatangelo, *Wound Repair Regen* 7, 79 (1999).
[3]   T. C. Laurent, U. B. Laurent and J. R. Fraser, *Ann. Med.* 28, 241 (1996).
[4]   L. Zhuo, V.C. Hascall and K. Kimata, *J. Biol. Chem.* 279, 38079 (2004).

[5] M. Yoneda, S. Suzuki and K. Kimata, *J. Biol. Chem.* 265, 5247 (1990).

[6] L. Huang, M. Yoneda and K. Kimata, *J. Biol. Chem.* 268, 26725 (1993).

[7] M. Zhao, M. Yoneda, Y. Ohashi, S. Kurono, H. Iwata, Y. Ohnuki and K. Kimata, *J. Biol. Chem.* 270, 26657 (1995).

[8] F. Bost, M. Diarra-Mehrpour and J. P. Martin, *Eur. J. Biochem.* 252, 339 (1998).

[9] L. Zhuo, M. Yoneda, M. Zhao, W. Yingsung, N. Yoshida, Y. Kitagawa, K. Kawamura, T. Suzuki and K. Kimata, *J. Biol. Chem.* 276, 7693 (2001).

[10] K. A. Hess, L. Chen, W. J. Larsen, *Biol. Reprod.* 61, 436 (1999).

[11] W. Yingsung, L. Zhuo, M. Morgelin, M. Yoneda, T. Kida, H. Watanabe, N. Ishiguro, H. Iwata and K. Kimata, *J. Biol. Chem.* 278, 32710 (2003).

[12] C. B. Basbaum and W. E. Finkbeiner, *Horm Metab Res* 20, 661 (1988).

[13] S. Sahu and W. S. Lynn, *Biochem. J.* 173, 565 (1978).

[14] T. S. Wilkinson, S. Potter-Perigo, C. Tsoi, L. C. Altman and T. N. Wight, *Am. J. Respir. Cell Mol. Biol.* 31, 92 (2004).

[15] M. E. Lauer, D. Mukhopadhyay, C. Fulop, C. A. de la Motte, A. K. Majors and V. C. Hascall, *J. Biol. Chem.* 284, 5299 (2008).

[16] M. Salathe, R. Forteza and G. E. Conner. *Novartis Found Symp.* 248, 20 (2002).

[17] L. Zhu, L. Zhuo, K Kimata, E. Yamaguchi, H. Watanabe, M. A. Aronica, V. C. Hascall and K. Baba, *Int Arch Allergy Immunol* 153, 223 (2010).

[18] P. S. Thomas, D. H. Yates and P. J. Barnes, *Am. J. Respir. Crit. Care Med.* 152, 76 (1995).

[19] M. Wong, D. Ziring, Y. KorinY, S. Desai, S. Kim, L. Lin, D. Gjertson, J. Braun, E. Reed and R. R. Singh, *Clin. Immunol.* 126, 121 (2008).

[20] N. Parameswaran and S. Patial, *Crit. Rev. Eukaryot Gene Expr.* 20, 87 (2010).

[21] J. B. Morjaria, A. J. Chauhan, K. S. Babu, R. Polosa, D. E. Davies and S. T. Holgate, *Thorax* 63, 584 (2008).

[22] M. K. Gately, D. M. Carvajal, S. E. Connaughton, S. Gillessen, R. R. Warrier, K. D. Kolinsky, V. L. Wilkinson, C. M. Dwyer, G. F. Higgins Jr., F. J. Podlaski, D. A. Faherty, P. C. Familletti, A. S. Stern and D.H. Presky, *Ann. N Y Acad. Sci.* 795, 1 (1996).

[23] F. Mattner, S. Fischer, S. Guckes, S. Jin, H. Kaulen, E. Schmitt, E. Rude and T. Germann, *Eur. J. Immunol.* 23, 2202 (1993).

[24] A. M. Cooper and S. A. Khader, *Trends Immunol.* 28, 33 (2007).

[25] D. E. Davies, *Proc. Am. Thorac. Soc.* 6, 678 (2009).

[26] B. R. Stripp, K. Maxson, R. Mera and G. Singh, *Am. J. Physiol. Lung Cell Mol Physiol* 269, L791 (1995).

[27] J. E. Adair, V. Stober, M. Sobhany, L. Zhuo, J. D. RobertsD, M. NegishiM, K. Kimata and S. Garantziotis, *J. Biol. Chem.* 284,16922 (2009).

[28] S. J. Wadsworth, A. M. Freyer, R. L. Corteling and I. P. Hall, *Am. J. Physiol. Lung Cell Mol. Physiol.* 286, L596 (2004).

[29] T. Miyazaki, M. Shen, D. Fujikura, N. Tosa, H. R. Kim, S. Kon, T. Uede and J. C. Reed, *J. Biol. Chem.* 279, 44667 (2004).

[30] M. H. Lazar, P. J. Christensen, M. Du, B. Yu, N. M. Subbotina, K. E. Hanson, J. M. Hansen, E. S. White, R. H. Simon and T. H. Sisson, *Am. J. Respir Cell Mol. Biol.* 31, 672 (2004).

[31] K. H. Minor, C. R. Schar, G. E. Blouse, J. D. Shore, D. A. Lawrence, P. Schuck and C. B. Peterson, *J. Biol. Chem.* 280, 28711 (2005).

[32] M. H. Siegelman, H. C. DeGrendele and P. Estess, *J. Leukoc Biol.* 66, 315 (1999).

[33] H. Ponta, L. Sherman and P. A. Herrlich. *Nat. Rev. Mol. Cell Biol.* 4, 33 (2003).

[34] L. Zhuo, A. Kanamori, R. Kannagi, N. Itano, J. Wu, M. Hamaguchi, N. Ishiguro and K. Kimata, *J. Biol. Chem.* 281, 20303 (2006).

[35] C. Galanos and M. A. Freudenberg, *Immunobiology* 187,346 (1993).

[36] C. Galanos, M. A. Freudenberg and W. Reutter, *Proc. Natl. Acad. Sci. USA* 76, 5939 (1979).

[37] A. Mignon, N. Rouque, M. Fabre, S. Martin, J. C. Pagès, J. F. Dhainaut, A. Kahn, P. Briand and V. Joulin, *Am. J. Respir. Crit. Care Med.* 159,1308 (1999).

[38] S. Q. Simpson and L. C. Casey, *Crit. Care Clin.* 5,27 (1989).

[39] M. Leist, F. Gantner, I. Bohlinger, P. G. Germann, G. Tiegs and A. Wendel, *J. Immunol.* 153,1778 (1994).

[40] Y. Wang, R. Singh, J. H. Lefkowitch, R. M. Rigol, and M. J. Czaja, *J. Biol. Chem.* 281, 15258 (2006).

[41] K. A. Brown, S. D. Brain, J. D. Pearson, J. D. Edgeworth, S. M. Lewis, and D. F. Treacher, *Lancet* 368, 157 (2006).

[42] J. F. Dhainaut, N. Marin, A. Mignon and C. Vinsonneau, *Crit. Care Med.* 29, S42 (2001).

[43] L. Liu and P. Kubes, *Thromb Haemost*, 89, 213 (2003).

[44] B. McDonald, E. F. McAvoy, F. Lam, V. Gill, C. de la Motte, R. C. Savani and P. Kubes, *J. Exp. Med.* 205, 915 (2008).

[45] B. McDonald, C. N. Jenne, L. Zhuo, K. Kimata and P. Kubes, *Am. J. Physiol. Gastrointest Liver Physiol.* 305, G797 (2013).

[46] A. Kanamori, N. Kojima, K. Uchimura, T. Muramatsu, T. Tamatani, M. C. Berndt, G. S. Kansas and R. Kannagi, *J. Biol. Chem.* 277, 32578 (2002).

[47] E. Shay, H. He, S. Sakurai, S. C. Tseng, *Invest Ophthalmol. Vis. Sci.* 52:2669 (2011).

In: Hyaluronan
Editor: Vitor H. Pomin

ISBN: 978-1-63117-808-5
© 2014 Nova Science Publishers, Inc.

Chapter 2

AGING AND HYALURONAN

Mamatha Damodarasamy and May J. Reed[*]
Division of Gerontology and Geriatric Medicine,
Department of Medicine
University of Washington School of Medicine, Seattle, US

ABSTRACT

The effect of aging on Hyaluronan (HA) has been studied primarily in the context of skin and cutaneous wound repair using animal models ranging from mice to human. Fetal skin contains high levels of higher molecular weight (HMW) HA, which is well described as inducing healing without fibrosis and scar formation. Adult skin is thought to retain HA content, but have alterations in HA synthesis. There is also evidence that aging alters cleavage of native HMW HA into lower MW forms. For example, we have found that age-related impairments in cutaneous wound repair are correlated with a deficit in the production of LMW-HA in aged mouse dermis. More recently, we have demonstrated that healing of dermal wounds in aged mice is significantly improved by treatment with HA of lower MW size. The influence of aging on HA content and MW in other tissues remains incompletely characterized. Aging might affect the extraction of HA, thereby altering the accurate measurement of HA content and size. Data in other organs such as the brain, prostate, and joints have evaluated HA content and size in the context of diseases that are more common with aging, rather than age

---

[*] PH: 206-897-5331, FAX: 206-744-9976, Email: mjr@uw.edu.

itself. These studies also suggest that with aging, most tissues modify HA content, size and influence on cellular processes relevant to injury, repair, and the development of pathology. This chapter will provide an overview of age-related changes in HA in the basal state and during injury in organs such as skin, brain, prostate and skeletal joints.

INTRODUCTION

Extracellular matrix (ECM) in connective tissue is comprised of numerous proteins that confer structure (such as the collagens) and regulatory activity (such as the matricellular glycoproteins). Both structural and regulatory ECM is highly associated with nonproteinaceous molecules, the most abundant of which is hyaluronan (HA). HA is a linear non-sulfated glycosaminoglycan (GAG) polymer of the disaccharide glucuronic acid/N-acetyl glucosamine that has a size range from $1->2 \times 10^4$ kDa. HA is highly hydrophilic and acts as a scaffold for organization of other ECM macromolecules, thereby mediating ECM assembly and homeostasis [1-3]. HA is extruded from the surfaces of cells by the action of HA synthases (HAS), of which HAS2 and -3 are primarily functional in the dermis [4, 5]. Whether specific HA synthases have effects on the size of HA produced is a matter of debate, but there is some evidence that HAS2 is of greater importance in maintaining HA content, HA size, and HA effects on tissue repair in skin [5-7]. Native HA can have a molecular weight (MW) as large as 2×10^4 kDa, but it is rapidly cleaved in the extracellular milieu into fragments ranging from 2–25,000 disaccharides by several mechanisms, including the activity of hyaluronidases (HYALases) [8, 9]. There are 6 HYALase-associated genes in humans, but the primary HYALases are HYALases 1-3. The presence of HYALase activity is important as HA has distinct biological effects that are determined, in part, by MW. Although this varies by cell type, forms of HA with a MW of 1×10^3 kDa and above ("high MW" HA) are generally thought to inhibit the proliferation and migration of cells. In contrast, lower MW (3–300 kDa) forms of HA usually promote cell proliferation and migration [8, 10-14]. These smaller HA forms are also pro-inflammatory, which can be beneficial or harmful depending on the organ and its physiologic state. For example, inciting inflammation in skin promotes endothelial cell proliferation and angiogenic pathways that could enhance cutaneous wound repair [15-17]. In the brain, HA is synthesized as a large molecule that is rapidly processed into LMW forms. This degradation is accelerated by inflammation and ischemia.

HA degradation results in LMW forms of HA that can further incite inflammation, thereby perpetuating the cycle [18, 19].

HA is unique among ECM molecules in that discrete sizes of HA can be formulated and administered safely in vivo [18, 19, Reed personal communication]. Regardless of the tissue, it is generally accepted that cells sense differences in HA size and responses reflect the size of the exogenous HA administered [20, 21].

HA AND AGING SKIN

Fetal Skin

The skin, especially the dermis, is rich in HA that is secreted by both keratinocytes and fibroblasts. Fetal skin contains higher HA content and greater expression of high MW HA as demonstrated by animal models ranging from mice to sheep [22-24]. These models generally support the premise that increased levels of high MW HA in fetal wounds promote healing without fibrosis and scar formation. The increase in HA content in fetal wounds when compared to adult wounds is attributed to reduced expression of HYALases [25]. Moreover, fetal fibroblasts demonstrate increased expression of the HA receptor CD44 relative to adult fibroblasts, indicating potential induction of HA utilization [26]. Reducing HA expression results in a phenotype more akin to adult healing in fetal rabbit wounds [27]. The effect of HA on scarless healing was further established in experiments with adult animal models. Incisional cutaneous wounds in adult rats treated with HA grafts (three dimensional HA strand modification) had reduced formation of scar tissue relative to controls who did not receive the addition of HA grafts [28, 29]. The mechanism by which high HA content affects scarring and fibrosis is likely an inhibitory effect on myofibroblast differentiation and subsequent suppression of the fibrogenic growth factor, Transforming Growth Factor-ß1 [30].

Not all effects of HA on cellular functions are inhibitory. Local levels of increased HA expression in developing tissues can also promote the proliferation and migration of a number of cell types. For example, in studies of embryonic development, hyaluronan-rich matrices occur around mesenchymal cells invading the primary corneal stroma; neural crest cells forming peripheral ganglia; somite cells encompassing developing vertebrae; cushion cells migrating to create heart valves; proliferating and migrating cells during brain development; and around mesenchymal cells during embryonic limb growth [31]. The role of HA as a promoter of tissue growth is also supported by its ubiquitous expression during appendage regeneration in amphibians, such as the newt [32]. In these scenarios, the extracellular matrices surrounding proliferating and migrating cells are highly enriched in HA [31]. A logical interpretation of this finding is that HA results in the hydration of tissues that promotes malleable matrices in which cells can readily replicate and migrate. In agreement with this idea are experiments such as one demonstrating that hyaluronan promotes glioblastoma cell migration within a fibrin gel by increasing hydration and consequently the porosity of the gel [33]. These mechanical concepts are separate from the consistent finding that HA can be processed into smaller forms that promote inflammatory, angiogenic and proliferative pathways during acute disruption of tissues. Depending on the stage of repair, these HA effects can either promote or inhibit wound healing [14, 16, 17, 25].

Aged Skin

In contrast to fetal skin, the influence of aging on HA content and MW in adult skin remains incompletely characterized [16, 34, 35]. This might reflect differences in extraction and organization with aging. We have found that the dermis of mice has similar total HA content and size with age [36]. There are certainly changes in HA in human skin due to both intrinsic (inherent) and extrinsic (environmental exposures, such as smoking and UV light) aging. Wrinkles and laxity are attributed, in part, to decreases in HA content [37, 38]. These findings have resulted in a prominent role for injectable HA in facial cosmetics. Changes in HA size with age are even less well characterized than that of HA content. There is evidence that in normal tissues cleavage of HA into lower MW forms is altered by aging, but no consistent patterns are noted [16, 39-41]. In our studies, HA size in normal dermis did not vary with age, although it is possible that different regions of dermis have distinct patterns of HA size that result in unique effects on inflammatory and repair processes.

Aging is characterized by deficits in multiple phases of dermal wound healing. These include impaired inflammatory responses, delayed cell proliferation, decreased synthesis of ECM components, disordered matrix deposition, impaired angiogenesis and an altered immunologic response [42]. Differences in structural matrix components, such as collagens, during wound repair in aging have long been well defined [43]. We have found that aging is associated with deficits in the ability to generate lower MW forms of HA during dermal wound repair [36]. Moreover, we have found that treatment of excisional wounds in aging with HA, especially HA of a middle size (250 kDa, HA250), results in improvement in dermal wound closure and is associated with significant increases in the synthesis and expression of collagen III (Reed, personal communication). There is biologic plausibility as to why HA250, which represents HA of intermediate size, would have the greatest benefit on wound repair – namely, this size range of HA does not trigger pathways inhibitory to wound repair as higher MW (>300kDa) forms of HA do, or provide the inflammatory signals in the manner of very low MW (<20kDa) forms of HA [11, 20, 21]. The ability of HA to promote collagen synthesis is in keeping with studies by others that have shown HA of a range of sizes can induce the synthesis of collagen III in a tissue specific manner [11, 44, 45]. Collagen III is often described as an early provisional collagen in tissue repair, but collagen III could also represent a marker for general increases in collagen content that are often associated with stimulation with HA. Indeed, HA effects on the synthesis of collagen III are mediated by fibrogenic growth factors in a manner similar to its influence on deposition of collagen I [46-48].

HA AND AGING IN OTHER ORGANS

Brain

Studies of HA in the brain have used a variety of models ranging from mice to humans. The widespread expression of HA in the brain was initially underappreciated due to losses of this carbohydrate during tissue processing. We now know that HA is synthesized primarily by astrocytes (a form of glial cell) in association with the lectins (chondroitin sulfate proteoglycans aggrecan, versican, neurocan, and brevican), the glycoprotein tenascin-R and the small protein, tumor necrosis factor-inducible gene 6 protein (TSG-6) [49, 50]. HA linkage to these and other components of brain ECM is mediated by proteins, such as Bral1, which confer additional stability. The expression of HA and HA-associated ECM molecules is noted throughout all the brain regions including cerebrum, cerebellum, and medulla [49-51]. Microscopically, HA and HA-associated ECM appear in densely packed areas of ECM called perineural nets. One specific portion of brain tissue that requires further study of HA is the blood brain barrier (BBB), which is comprised of endothelial cells and pericytes that regulate passage of substances from the systemic circulation into the brain [52]. In conjunction with its abundant expression on the abluminal (ECM) side of the BBB, HA is the dominant component of the endothelial glycocalyx lining the vascular (luminal) side of the BBB. The BBB is more prone to disruption with age, but whether HA thickness or size in the glycocalyx contributes to this vulnerability is unknown. HA and HA-associated ECM expression, synthesis, turnover and degradation are noted as neural tissue matures and are often found in areas of brain ischemia and scar formation [53]. These data have led to the concept that brain ECM is integral to neurotransmitter diffusion, synaptic function, and activation of signaling pathways [51]. HA expression increases in areas of inflammation or ischemia where it is thought to both inhibit oligodendrocyte precursor cell maturation and modulate signals that can potentiate additional neuronal injury [53-55].

Little is known about HA and normal brain aging. Utilizing electrophoretic separation of glycosaminoglycans, a substantial increase in HA relative to chondroitin sulfate concentration was found in the brain tissue of 30 month-old rats compared to tissues from younger rats ranging from postnatal day 10 to 18 months of age [56]. More recently, HA content, as measured by a highly sensitive enzyme-linked immunosorbent assay (ELISA)-based assay, has been reported to increase with normal aging, especially in general brain gray matter [57]. Reports that HA expression might decrease in specific regions of the brain with aging usually compare adult brains with developing brains that are known to be abundant in HA. Even in these comparisons, which do not include an old group of animals, it is notable that high expression of HA is retained in regions of the adult brain that are rich in neural stem cells and progenitor cells, such as the rostral migratory stream and the subventricular zone of mouse brains [58]. No studies to date have found specific differences in regional expression of HA or its associated ECM proteins in the brain with normal aging.

Clinically, deposition of HA is found in areas of vascular brain injury that appear as white matter hyperintensities, which represent increased density of immunoreactive cells, on diffusion tensor imaging (a form of MRI) [59]. Interestingly, HA expression is not found in areas of plaques related to Alzheimer's pathology. There are significant gender differences in HA content in the cerebral spinal fluid (CSF) of patients with Alzheimer's disease, with affected males having twice as much HA in their CSF as affected females. In addition, the CSF of females with AD demonstrated correlations between HA levels, AD-related biomarkers and inflammatory biomarkers [60]. Whether the appearance of HA and its associated ECM is ultimately beneficial or detrimental to neural plasticity and repair is a matter of ongoing debate [61, 62]. The sizes of HA that mediate specific functions in the CNS are also an area of active investigation. For example, injection of HA of distinct sizes and modulation of HA size (via HYALases) into the peripheral circulation (by subcutaneous injection) as well as directly into the CNS (via injection into the rostral corpus callosum) can affect ongoing inflammatory processes in experimental models of encephalitis and multiple sclerosis, primarily by interaction with the receptor Toll-like receptors 2 and 4 (TLR2 and TLR4) [18, 19, 55].

Prostate

HA is highly expressed by stromal cells in the prostate interstitium, the ECM-rich regions between the secretory glands that produce prostatic fluid, in models ranging from mice to humans. HA and HA-associated ECM components are found in increased amounts in human prostate ECM with both benign (prostatic hyperplasia) and malignant (cancer) pathology [63, 64]. Accordingly, there is a long-standing interest in HA content, HA size, and levels of mediators of HA cleavage in prostate [10, 65, 66]. As in other organs, HA size has differential effects on prostate tissues that relate to its chain length: newly-synthesized HA has a high MW (up to 2×10^4 kDa) and can inhibit a variety of cellular processes [67, 68]. Once secreted, however, HA is rapidly cleaved by hyaluronidases into a continuum of lower MW forms (i.e., < 300 kDa) that can promote the growth of prostate tissue. Consequently, the role of HA content, size, and mediators of HA cleavage as predictors for disease progression in the prostate is a subject of intense study [10, 65, 66, 69, 70]. In prostate cancer, it has been shown that increased HA content, accompanied by increased HYALase levels and elevated HA degradation, correlates with aggressive disease [10, 65, 66, 69-72]. HYALase levels in urine are being evaluated as potential biomarkers for prostate cancer progression [73]. Blockade of HA receptors for endocytosis (HARE) on endothelial cells represents a novel therapeutic target to prevent the metastasis of prostate cancers [74]. Far less is known about the influence of HA on benign prostate hyperplasia, but it is generally accepted that an increase in total HA expression and variability of HA size is a feature of benign prostate disease [63, 64].

The potential role of HA content and size in cancer progression has recently been highlighted by the naked mole rat, which is a long-lived mammal that remains cancer free [75]. In this model there is high HA content and an abundance of high molecular weight HA. The latter is thought to be due to an increase in HAS2 expression and decrease in Hyaluronidase activity. The increase in HA content is seen in multiple tissues including skin, heart, brain, and kidney. These results have supported the concept that the naked mole rat is a valuable model for the study of biochemical and molecular traits that confer both longevity and resistance to cancer formation.

Joints

High MW HA is typically found in abundance in the synovial fluid of human joints where it is tightly associated with the sulfated GAG, chondroitin sulfate, and the mucinous glycoprotein, lubricin [76, 77]. The primary cell that secretes HA is the synovial cell, a specialized type of secretory fibroblast that lines the joint space. It is generally accepted that age-related deficits in HA content in these regions contribute to degenerative joint disease by exacerbating concurrent losses of proteoglycan and collagen content (in cartilage and bone) as well as overall hydration of the joint space. Factors such as age, weight and sex affect the metabolism of chondroitin sulfate and HA and contribute to the onset and severity of arthritis. Synovial fluid obtained from different ages (16-89) showed a significant negative correlation of HA and chondroitin sulphate with aging [78]. Serum HA level increased with the severity of knee osteoarthritis [79]. Based on these data, HA derived from rooster combs has long been available for clinical use in joints affected by osteoarthritis via a technique called viscosupplementation. There are numerous studies on intra articular HA therapy for osteoarthritis and hip treatment [80-82]. Most use a form of HA that is > 500 kDa and support the concept that injection of HA relieves pain and improves function of both small and large joints. Others propose that viscosupplementation with HA actually improves joint lubrication and delays the progression of osteoarthritis [81]. Recent reviews of randomized trials with appropriate controls (sham or saline) have been conflicting. One showed a small, but clinically irrelevant, benefit accompanied by a greater risk of adverse side effects and the other demonstrated a benefit of HA in improving both pain and function in the knee joints [83, 84]. Further studies are needed to determine the duration of benefit, if any, and whether there are long term sequelae associated with repeated injections of HA.

CONCLUSION

Hyaluronan is a ubiquitous component of the ECM. Its notable effects early in development are related to its ability to affect scarless tissue repair. As animals age, the expression of HA is still widespread, but its role is determined by its association with other ECM components (such as collagens) and its MW. In all tissues (except possibly the brain), HA is continuously synthesized and degraded. It is generally accepted that the persistence of larger forms of HA can inhibit proliferative activities, thereby decreasing the development of cancers and possibly inflammation. Conversely, the formation of middle and lower MW sizes of HA can affect immune responses that can be beneficial in tissue repair, but are potentially deleterious in the setting of brain injury or cancer formation.

ACKNOWLEDGMENTS

NIH: R21AG33391

REFERENCES

[1]   Evanko, S.P., et al., Hyaluronan-dependent pericellular matrix. *Adv. Drug Deliv. Rev.*, 2007. 59(13): p. 1351-65.

[2]   Roughley, P.J., et al., The role of hyaluronan produced by Has2 gene expression in development of the spine. *Spine* (Phila Pa 1976), 2011. 36(14): p. E914-20.

[3]   Toole, B.P., Hyaluronan: from extracellular glue to pericellular cue. *Nat. Rev. Cancer*, 2004. 4(7): p. 528-39.

[4]   Averbeck, M., et al., Differential regulation of hyaluronan metabolism in the epidermal and dermal compartments of human skin by UVB irradiation. *J. Invest. Dermatol.*, 2007. 127(3): p. 687-97.

[5]   Gebhardt, C., et al., Dermal hyaluronan is rapidly reduced by topical treatment with glucocorticoids. *J. Invest. Dermatol.*, 2010. 130(1): p. 141-9.

[6]   Mack, J.A., et al., Enhanced inflammation and accelerated wound closure following tetraphorbol ester application or full-thickness

wounding in mice lacking hyaluronan synthases Has1 and Has3. *J. Invest. Dermatol.*, 2012. 132(1): p. 198-207.

[7] Oh, J.H., et al., Changes in glycosaminoglycans and related proteoglycans in intrinsically aged human skin in vivo. *Exp. Dermatol.*, 2011. 20(5): p. 454-6.

[8] Stern, R., Hyaluronan metabolism: a major paradox in cancer biology. *Pathol. Biol.* (Paris), 2005. 53(7): p. 372-82.

[9] Toole, B.P., S. Ghatak, and S. Misra, Hyaluronan oligosaccharides as a potential anticancer therapeutic. *Curr. Pharm. Biotechnol.*, 2008. 9(4): p. 249-52.

[10] Bharadwaj, A.G., K. Rector, and M.A. Simpson, Inducible hyaluronan production reveals differential effects on prostate tumor cell growth and tumor angiogenesis. *J. Biol. Chem.*, 2007. 282(28): p. 20561-72.

[11] David-Raoudi, M., et al., Differential effects of hyaluronan and its fragments on fibroblasts: relation to wound healing. *Wound Repair Regen.*, 2008. 16(2): p. 274-87.

[12] Jiang, D., J. Liang, and P.W. Noble, Hyaluronan as an immune regulator in human diseases. *Physiol. Rev.*, 2011. 91(1): p. 221-64.

[13] Moon, S.O., J.H. Lee, and T.J. Kim, Changes in the expression of c-myc, RB and tyrosine-phosphorylated proteins during proliferation of NIH 3T3 cells induced by hyaluronic acid. *Exp. Mol. Med.*, 1998. 30(1): p. 29-33.

[14] Slevin, M., S. Kumar, and J. Gaffney, Angiogenic oligosaccharides of hyaluronan induce multiple signaling pathways affecting vascular endothelial cell mitogenic and wound healing responses. *J. Biol. Chem.*, 2002. 277(43): p. 41046-59.

[15] Montesano, R., et al., Synergistic effect of hyaluronan oligosaccharides and vascular endothelial growth factor on angiogenesis in vitro. *Lab. Invest.*, 1996. 75(2): p. 249-62.

[16] Stern, R. and H.I. Maibach, Hyaluronan in skin: aspects of aging and its pharmacologic modulation. *Clin. Dermatol.*, 2008. 26(2): p. 106-22.

[17] West, D.C. and S. Kumar, The effect of hyaluronate and its oligosaccharides on endothelial cell proliferation and monolayer integrity. *Exp. Cell Res.*, 1989. 183(1): p. 179-96.

[18] Winkler, C.W., et al., Hyaluronan oligosaccharides perturb lymphocyte slow rolling on brain vascular endothelial cells: implications for inflammatory demyelinating disease. *Matrix Biol.*, 2013. 32(3-4): p. 160-8.

[19] Winkler, C.W., et al., Hyaluronan anchored to activated CD44 on central nervous system vascular endothelial cells promotes lymphocyte extravasation in experimental autoimmune encephalomyelitis. *J. Biol. Chem.*, 2012. 287(40): p. 33237-51.

[20] Ghazi, K., et al., Hyaluronan fragments improve wound healing on in vitro cutaneous model through P2X7 purinoreceptor basal activation: role of molecular weight. *PLoS One*, 2012. 7(11): p. e48351.

[21] Maharjan, A.S., D. Pilling, and R.H. Gomer, High and low molecular weight hyaluronic acid differentially regulate human fibrocyte differentiation. *PLoS One,* 2011. 6(10): p. e26078.

[22] Longaker, M.T., et al., Studies in fetal wound healing. V. A prolonged presence of hyaluronic acid characterizes fetal wound fluid. *Ann. Surg.,* 1991. 213(4): p. 292-6.

[23] Rolfe, K.J. and A.O. Grobbelaar, A review of fetal scarless healing. *ISRN Dermatol.*, 2012. 2012: p. 698034.

[24] Sawai, T., et al., Hyaluronic acid of wound fluid in adult and fetal rabbits. *J. Pediatr. Surg.*, 1997. 32(1): p. 41-3.

[25] West, D.C., et al., Fibrotic healing of adult and late gestation fetal wounds correlates with increased hyaluronidase activity and removal of hyaluronan. *Int. J. Biochem. Cell Biol.*, 1997. 29(1): p. 201-10.

[26] Alaish, S.M., et al., Biology of fetal wound healing: hyaluronate receptor expression in fetal fibroblasts. *J. Pediatr. Surg.*, 1994. 29(8): p. 1040-3.

[27] Mast, B.A., et al., Hyaluronic acid modulates proliferation, collagen and protein synthesis of cultured fetal fibroblasts. *Matrix*, 1993. 13(6): p. 441-6.

[28] Hu, M., et al., Three-dimensional hyaluronic acid grafts promote healing and reduce scar formation in skin incision wounds. *J. Biomed. Mater. Res. B. Appl. Biomater.*, 2003. 67(1): p. 586-92.

[29] Iocono, J.A., et al., Hyaluronan induces scarless repair in mouse limb organ culture. *J. Pediatr. Surg.*, 1998. 33(4): p. 564-7.

[30] Penn, J.W., A.O. Grobbelaar, and K.J. Rolfe, The role of the TGF-beta family in wound healing, burns and scarring: a review. *Int. J. Burns Trauma,* 2012. 2(1): p. 18-28.

[31] Toole, B.P., Hyaluronan in morphogenesis. *Semin. Cell Dev. Biol.*, 2001. 12(2): p. 79-87.

[32] Calve, S., S.J. Odelberg, and H.G. Simon, A transitional extracellular matrix instructs cell behavior during muscle regeneration. *Dev. Biol.,* 2010. 344(1): p. 259-71.

[33] Hayen, W., et al., Hyaluronan stimulates tumor cell migration by modulating the fibrin fiber architecture. *J. Cell Sci.*, 1999. 112 (Pt 13): p. 2241-51.

[34] Ghersetich, I., et al., Hyaluronic acid in cutaneous intrinsic aging. *Int. J. Dermatol.*, 1994. 33(2): p. 119-22.

[35] Manuskiatti, W. and H.I. Maibach, Hyaluronic acid and skin: wound healing and aging. *Int. J. Dermatol.*, 1996. 35(8): p. 539-44.

[36] Reed, M.J., et al., Cleavage of hyaluronan is impaired in aged dermal wounds. *Matrix Biol.*, 2013. 32(1): p. 45-51.

[37] Fisher, G.J., et al., Collagen fragmentation promotes oxidative stress and elevates matrix metalloproteinase-1 in fibroblasts in aged human skin. *Am. J. Pathol.*, 2009. 174(1): p. 101-14.

[38] Quan, T., et al., Enhancing structural support of the dermal microenvironment activates fibroblasts, endothelial cells, and keratinocytes in aged human skin in vivo. *J. Invest. Dermatol.*, 2013. 133(3): p. 658-67.

[39] Meyer, L.J. and R. Stern, Age-dependent changes of hyaluronan in human skin. *J. Invest. Dermatol.*, 1994. 102(3): p. 385-9.

[40] Miyamoto, I. and S. Nagase, Age-related changes in the molecular weight of hyaluronic acid from rat skin. *Jikken Dobutsu*, 1984. 33(4): p. 481-5.

[41] Robert, L., A.M. Robert, and G. Renard, Biological effects of hyaluronan in connective tissues, eye, skin, venous wall. Role in aging. *Pathol. Biol.* (Paris), 2010. 58(3): p. 187-98.

[42] Sgonc, R. and J. Gruber, Age-related aspects of cutaneous wound healing: a mini-review. *Gerontology*, 2013. 59(2): p. 159-64.

[43] Ashcroft, G.S., et al., Estrogen accelerates cutaneous wound healing associated with an increase in TGF-beta1 levels. *Nat. Med.*, 1997. 3(11): p. 1209-15.

[44] Li, Y., et al., Irradiation-induced expression of hyaluronan (HA) synthase 2 and hyaluronidase 2 genes in rat lung tissue accompanies active turnover of HA and induction of types I and III collagen gene expression. *Am. J. Respir. Cell Mol. Biol.*, 2000. 23(3): p. 411-8.

[45] Wang, F., et al., In vivo stimulation of de novo collagen production caused by cross-linked hyaluronic acid dermal filler injections in photodamaged human skin. *Arch. Dermatol.*, 2007. 143(2): p. 155-63.

[46] Appling, W.D., et al., Synergistic enhancement of type I and III collagen production in cultured fibroblasts by transforming growth factor-beta and ascorbate. *FEBS Lett.*, 1989. 250(2): p. 541-4.

[47] Varga, J., J. Rosenbloom, and S.A. Jimenez, Transforming growth factor beta (TGF beta) causes a persistent increase in steady-state amounts of type I and type III collagen and fibronectin mRNAs in normal human dermal fibroblasts. *Biochem. J.,* 1987. 247(3): p. 597-604.

[48] Wang, X., et al., Effects of TRAP-1-like protein (TLP) gene on collagen synthesis induced by TGF-beta/Smad signaling in human dermal fibroblasts. *PLoS One,* 2013. 8(2): p. e55899.

[49] Zimmermann, D.R. and M.T. Dours-Zimmermann, Extracellular matrix of the central nervous system: from neglect to challenge. *Histochem. Cell Biol.,* 2008. 130(4): p. 635-53.

[50] Higman, V.A., et al., A Refined Model for the TSG-6 Link Module in Complex with Hyaluronan: Use of defined oligosaccharides to probe structure and function. *J. Biol. Chem.,* 2014. 289(9): p. 5619-34.

[51] Howell, M.D. and P.E. Gottschall, Lectican proteoglycans, their cleaving metalloproteinases, and plasticity in the central nervous system extracellular microenvironment. *Neuroscience,* 2012. 217: p. 6-18.

[52] Daneman, R., The blood-brain barrier in health and disease. *Ann. Neurol.,* 2012. 72(5): p. 648-72.

[53] Al'Qteishat, A., et al., Changes in hyaluronan production and metabolism following ischaemic stroke in man. *Brain,* 2006. 129(Pt 8): p. 2158-76.

[54] Back, S.A., et al., Hyaluronan accumulates in demyelinated lesions and inhibits oligodendrocyte progenitor maturation. *Nat. Med.,* 2005. 11(9): p. 966-72.

[55] Sloane, J.A., et al., Hyaluronan blocks oligodendrocyte progenitor maturation and remyelination through TLR2. *Proc. Natl. Acad. Sci. U S A,* 2010. 107(25): p. 11555-60.

[56] Jenkins, H.G. and H.S. Bachelard, Developmental and age-related changes in rat brain glycosaminoglycans. *J. Neurochem.,* 1988. 51(5): p. 1634-40.

[57] Cargill, R., et al., Astrocytes in aged nonhuman primate brain gray matter synthesize excess hyaluronan. *Neurobiol. Aging,* 2012. 33(4): p. 830 e13-24.

[58] Lindwall, C., et al., Selective expression of hyaluronan and receptor for hyaluronan mediated motility (Rhamm) in the adult mouse subventricular zone and rostral migratory stream and in ischemic cortex. *Brain Res.,* 2013. 1503: p. 62-77.

[59] Back, S.A., et al., White matter lesions defined by diffusion tensor imaging in older adults. *Ann. Neurol.,* 2011. 70(3): p. 465-76.

[60] Nielsen, H.M., et al., Gender-dependent levels of hyaluronic acid in cerebrospinal fluid of patients with neurodegenerative dementia. *Curr Alzheimer. Res., 2012.* 9(3): p. 257-66.

[61] Gaal, B., et al., Distribution of extracellular matrix macromolecules in the vestibular nuclei and cerebellum of the frog, *Rana esculenta. Neuroscience*, 2014. 258: p. 162-73.

[62] Frischknecht, R. and E.D. Gundelfinger, The brain's extracellular matrix and its role in synaptic plasticity. *Adv. Exp. Med. Biol.*, 2012. 970: p. 153-71.

[63] De Klerk, D.P., The glycosaminoglycans of normal and hyperplastic prostate. *Prostate,* 1983. 4(1): p. 73-81.

[64] Goulas, A., et al., Benign hyperplasia of the human prostate is associated with tissue enrichment in chondroitin sulphate of wide size distribution. *Prostate,* 2000. 44(2): p. 104-10.

[65] Kovar, J.L., et al., Hyaluronidase expression induces prostate tumor metastasis in an orthotopic mouse model. *Am. J. Pathol.*, 2006. 169(4): p. 1415-26.

[66] Lokeshwar, V.B., et al., Stromal and epithelial expression of tumor markers hyaluronic acid and HYAL1 hyaluronidase in prostate cancer. *J. Biol. Chem.*, 2001. 276(15): p. 11922-32.

[67] Bharadwaj, A.G., et al., Hyaluronan suppresses prostate tumor cell proliferation through diminished expression of N-cadherin and aberrant growth factor receptor signaling. *Exp. Cell Res.*, 2011. 317(8): p. 1214-25.

[68] Goldberg, R.L. and B.P. Toole, Hyaluronate inhibition of cell proliferation. *Arthritis Rheum.*, 1987. 30(7): p. 769-78.

[69] Bharadwaj, A.G., et al., Spontaneous metastasis of prostate cancer is promoted by excess hyaluronan synthesis and processing. *Am. J. Pathol.*, 2009. 174(3): p. 1027-36.

[70] Lokeshwar, V.B., et al., Association of elevated levels of hyaluronidase, a matrix-degrading enzyme, with prostate cancer progression. *Cancer Res,* 1996. 56(3): p. 651-7.

[71] Misra, S., B.P. Toole, and S. Ghatak, Hyaluronan constitutively regulates activation of multiple receptor tyrosine kinases in epithelial and carcinoma cells. *J. Biol. Chem.*, 2006. 281(46): p. 34936-41.

[72] Simpson, M.A. and V.B. Lokeshwar, Hyaluronan and hyaluronidase in genitourinary tumors. *Front Biosci.*, 2008. 13: p. 5664-80.

[73] Chib, R., et al., FRET based Ratio-metric Sensing of Hyaluronidase in Synthetic Urine as a Biomarker for Bladder and Prostate Cancer. *Curr. Pharm. Biotechnol.*, 2013.

[74] Simpson, M.A., J.A. Weigel, and P.H. Weigel, Systemic blockade of the hyaluronan receptor for endocytosis prevents lymph node metastasis of prostate cancer. *Int. J. Cancer*, 2012. 131(5): p. E836-40.

[75] Tian, X., et al., High-molecular-mass hyaluronan mediates the cancer resistance of the naked mole rat. *Nature*, 2013. 499(7458): p. 346-9.

[76] Hui, A.Y., et al., A systems biology approach to synovial joint lubrication in health, injury, and disease. *Wiley Interdiscip. Rev. Syst. Biol. Med.*, 2012. 4(1): p. 15-37.

[77] Murano, E., et al., Hyaluronan: from biomimetic to industrial business strategy. *Nat. Prod. Commun.*, 2011. 6(4): p. 555-72.

[78] Uesaka, S., K. Miyazaki, and H. Ito, Age-related changes and sex differences in chondroitin sulfate isomers and hyaluronic acid in normal synovial fluid. *Mod. Rheumatol.*, 2004. 14(6): p. 470-475.

[79] Inoue, R., et al., Knee osteoarthritis, knee joint pain and aging in relation to increasing serum hyaluronan level in the Japanese population. *Osteoarthritis Cartilage*, 2011. 19(1): p. 51-7.

[80] Migliore, A. and M. Granata, Intra-articular use of hyaluronic acid in the treatment of osteoarthritis. *Clin. Interv. Aging*, 2008. 3(2): p. 365-9.

[81] Elmorsy, S., et al., Chondroprotective effects of high-molecular-weight cross-linked hyaluronic acid in a rabbit knee osteoarthritis model. *Osteoarthritis Cartilage*, 2013.

[82] Gigante, A. and L. Callegari, The role of intra-articular hyaluronan (Sinovial) in the treatment of osteoarthritis. *Rheumatol. Int.*, 2011. 31(4): p. 427-44.

[83] Miller, L.E. and J.E. Block, US-Approved Intra-Articular Hyaluronic Acid Injections are Safe and Effective in Patients with Knee Osteoarthritis: Systematic Review and Meta-Analysis of Randomized, Saline-Controlled Trials. *Clin. Med. Insights Arthritis Musculoskelet. Disord.*, 2013. 6: p. 57-63.

[84] Rutjes, A.W., et al., Viscosupplementation for osteoarthritis of the knee: a systematic review and meta-analysis. *Ann. Intern. Med.*, 2012. 157(3): p. 180-91.

In: Hyaluronan   ISBN: 978-1-63117-808-5
Editor: Vitor H. Pomin   © 2014 Nova Science Publishers, Inc.

Chapter 3

BIOLOGICAL SYNTHESIS AND CATABOLISM OF HYALURONIC ACID DURING CARCINOGENESIS

Rajeev K. Boregowda[1],* and Shib Das Banerjee[2]
[1]Department of Medicine, Medical Oncology, Rutgers Cancer
Institute of New Jersey, Rutgers, The State University, NJ, US
[2]Department of Anatomy, Tuft University
School of Medicine, Boston, MA, US

ABSTRACT

Most of the cells in multicellular organisms are in contact with an intricate meshwork of interacting, extracellular macromolecules that constitute an extracellular matrix (ECM). The ECM plays an important role, not only in tissue architecture but also in the regulation of gene expression and cell behavior. Many of the ECM effects are mediated by dynamic interaction between ECM macromolecules and their receptors.

Among the ECM components, Hyaluronic Acid (HA) is one of the most fascinating molecules. HA is a high molecular weight, uniformly repetitive, and linear glycosaminoglycan (GAG), composed of disaccharides of glucuronic acid and N-acetylglucosamine.

HA is large; not only in molecular weight, but also in the space it occupies in solution, and its versatile biological function is completely

---

*E-mail address: rajeev04mys@yahoo.com

dependent on its size. Earlier studies showed that cells divide and migrate within ECM rich in HA. The dramatic modulation of HA concentration and organization accompany cellular changes that take place during tissue and organ differentiation. In recent years, ample studies have revealed the critical role of HA in carcinogenesis in different types of cancer suggesting that it can be an independent prognostic indicator. During tumor development, cancerous cells tend to synthesize more HA, which accumulates in the intracellular compartments and pericellular matrices.

The accumulated HA in the tumor is known to activate signaling pathways that are favorable for tumor growth and progression. These effects are mediated through the interaction between HA and its binding proteins (HABPs, Hyaluronic Acid Binding Proteins). Further during tumor progression, catabolism of HA to HA oligosaccharides by hyaluronidase enzyme leads to angiogenesis and activation of a new array of cell-signaling events. Therefore, targeting aberrant HA synthesis and accumulation could potentially prevent tumor growth.

In this chapter, we will discuss the biological synthesis of hyaluronic acid during normal developmental processes and carcinogenesis. Further, we will discuss the catabolism of HA to HA oligosaccharides. Finally, we will discuss the possible ways to prevent aberrant HA synthesis and its accumulation during carcinogenesis.

Keywords: hyaluronic acid, malignant tumor, targeting hyaluronic acid, hyaluronic acid binding protein, hyaluronic acid catabolism, hyaluronic acid biosynthesis

INTRODUCTION

The ECM is an integral part of tissue microenvironments, and it is no longer considered as an inert substance because it initiates crucial biochemical and biomechanical signals during tissue morphogenesis, differentiation and homeostasis [1]. The ECM is made of an interlaced network of fibrous protein such as collagens, elastins, fibronectin, laminins [2], and glycosaminoglycans (GAGs). The GAGs, except hyaluronic acid, are covalently attached to a specific protein core to form proteoglycans (PGs) [3, 4]. The three main families of PGs are cell surface, modular, and small leucine-rich proteoglycans [4, 5]. The GAGs in the PG are unbranched polysaccharides chains, consisting of repeating disaccharide units and made up of sulfated (chondroitin sulfate, heparin sulfate, and keratin sulfate) and non-sulfated GAGs such as hyaluronic acid [4]. As such, PGs have a net negative charge. Thus, they are extremely hydrophilic and can hold a large amount of water. This property allows the tissue to withstand high compressive pressure.

Therefore, the tissue ECM is highly dynamic in terms of its molecular composition and its biological functions, and that must be properly orchestrated for the maintenance of normal function [6].

Therefore, there is an explosion in research studies with aim to utilize ECM as scaffolds in various medical applications such as wound healing, bone grafting and whole-organ tissue engineering [7-10].

1. ROLE OF HA DURING EARLY DEVELOPMENTAL PROCESS

Hyaluornic acid (also known as Hyaluronan) was first isolated in 1935 from the vitreous body of cows' eyes by Karl Meyer [11], and subsequently its structure was resolved by Meyer and his co-workers [12].

According to them, HA is high-molecular-weight, linear polysaccharide with repeating disaccharide units containing [D-Glucuronic acid (1-β-3) N-acetyl-D-Glucosamine (1-β-4)] $_n$ units, and in a full length HA the number of disaccharides units can be 10000 or more with a molecular weight of $\sim 4 \times 10^6$ (4 million Daltons). Under physiological conditions the HA acquire a stable helical configuration stabilized by hydrogen bonds which run parallel along the chain axis, resulting in extensive hydrophobic patches [13]. As a result, HA can hold a large amount of water relative to its weight though it is not chemically bound to the HA [14, 15] and, at higher concentration, the HA solution is highly viscous and can withstand mechanical compression, similar to the properties of a liquid lubricant [14, 16]. Because of these properties, the HA molecule has been used in a wide variety of medical and pharmacological applications [16, 17].

HA is not synthesized at the Golgi complex, in contrast to other glycasminonoglycans. Instead, HA is synthesized inside the cell membrane by adding monosaccharides to the reducing end of the chain, and then translocated to pericellular space [18]. There are three HA synthases (HASs), named HAS1, HAS2 and HAS3. These are isoforms, and are responsible for the biosynthesis of HA [18, 19]. Among these HAS enzymes, HAS2 has been shown to be crucial for HA biosynthesis during the developmental process [20].

HA is an ancient molecule, present in bacteria as well as in vertebrates [21-23], and its critical role in the development is evidenced by its presence throughout the body at varying concentrations [23]. The biological synthesis of HA takes place during early embryonic development, and has been shown to occur around the sixth week of gestation in the vitreous during embryonic development [24], limb development [20, 25], and early development of the brain [26]. Histochemical experiments have shown that newly forming blood capillaries tips are enriched in HA [27]. Further, it has been shown that HA is a vital component of the stem cell niche [28, 29], lending evidence of its role in early development. Many studies have shown that HA synthesis peaks at mitosis during cell proliferation, and increased HA content would help the cells detach from the matrix to initiate the cell division [30].

However, HA synthesis declines once the cells progress towards differentiation process [31, 32]. In addition to extracellular HA, there are research reports proving the presence of intracellular HA in the cytoplasm, nucleus and other organelles [33-36].

The intracellular HA is thought to participate in important cellular processes, such as cell division, by facilitating the nuclei separation process (see review [37]) and growth plate expansion during bone development [38]. HA localization has been observed in several organelles such as cytoplasm, mitochondria, endoplasmic reticulum, and the nucleus [37] (also see [39]).

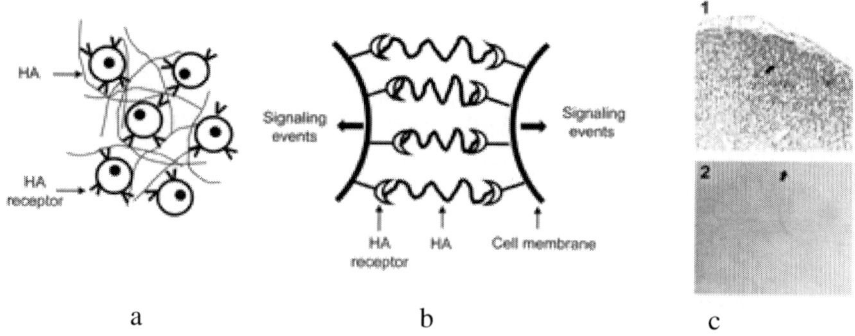

Figure 1. Biological synthesis of HA during the developmental process. A) During early developmental process, the presence of excess of HA creates loose environment to promote cell division. B) Later, during the differentiation process, HA is shown to attach to its receptor on the cell surface to promote cell-to-cell communication. C1) Section of the chick embryo limb showing increased accumulation of HA (indicated by the arrow) in the developing limb bud, detected by using biotinylated PG. C2) Chick embryo limb section stained without biotinylated PG.

However, it is not clear how much of the intracellular HA originated from re-internalization of HA from outside the cell through the endocytosis process, and how it reaches different sites in the cell.

Understanding the above mentioned biological functions of HA raised the question of how these effects are mediated.

To this end, previous studies identified several HA binding proteins, such as CD44 [40], the receptor for HA-mediated motility (RHAMM) [41]; lymphatic vessel endothelial HA receptor (LYVE)-1 [42], the HA receptor for endocytosis; HARE/Stabilin-2 [43-45]; Toll-like receptors-2 and 4 [46],; and H11B2C2 mAb reactive HABP cell-surface receptors (this HABP was also observed in cytoplasm) [47]. Conversely, CDC37 [48], P32 [49], intracellular HA binding protein (IHABP) -4 [39], and RHAMM/IHABP [50, 51] are intra-cellular receptors for HA.

2. SYNTHESIS AND CATABOLISM OF HA DURING CARCINOGENESIS

2.1. Aberrant Synthesis of HA and Its Effect on Carcinogenesis

Cancer is characterized by the loss of tissue framework and the aberrant behavior of cells. These events are linked to genetic mutations and epigenetic modifications. The ECM of the tumor is very similar to the ECM found in an unresolved wound [52].

Since there are commonalities between normal and cancer developmental processes and given the role of HA in normal development, it is conceivable that HA might also play an important role in carcinogenesis.

In support of this idea, there is enough evidence available to prove the role HA plays in many cancer types [53-56] (also, see excellent reviews published in Seminars in Cancer Biology, volume 18, 2008). Several studies have shown the relationship between HA synthesis and tumor progression in cell culture, in both animal models and human tumor samples.

The increased HA synthesis during tumor development is produced by either tumor cells or the cells in the tumor microenvironment.

Detailed histochemical studies using human tumor samples have confirmed high levels of HA in the tumor cells, as well as in the tumor micro-environment [53, 55, 57, 58] (also, see review [59]), and its association with tumor progression.

In addition, it has been shown that high HA accumulation in the cytoplasm is significantly associated with unfavorable outcomes in colorectal, gastric, and breast cancer [60-62]. During carcinogenesis, differential transcriptional regulation that leads to selective expression of HA synthase (HAS) isoforms has been implicated for increased HA synthesis [63]. In support of this finding, overexpression of the HAS gene has been shown to increase the tumorigenic ability of mesothelioma cells [64], prostate cancer cells [65], and melanoma cells [66]. Conversely, HAS2 overexpression in glioma cells suppressed their tumorigenesis [67]. These findings raise the question of why tumor cells, depending on the type, selectively express specific form of HAS, and the nature of the mechanism behind this regulation. During tumor progression, HA is known to induce tumor growth by stimulating anchorage-independent growth [68]. In addition, the biophysical property of HA also creates the perfect environment for the survival, growth, and invasion of tumor cells. To support this notion, HA accumulation during carcinogenesis is known to impair cell-to-cell contact inhibition to promote tumor cell growth and migration [69]. There are striking similarities between wound healing and cancer. However, the healing process is not self-controlled in cancer and results in malignant tumor development, invasion, and metastasis. In 1863, Rudolf Virchow postulated that chronic wounds can lead to tumorigenesis [70]. Some of the earlier studies of inflammation caused by chronic viral hepatitis and Helicobacter Pylori infection, and inflammation bowl disease [71], revealed the similarities in cellular and molecular mechanisms between wound and cancer tissue. In fact, it was postulated that "tumors are wounds that do not heal" [72]. The role of increased inflammation in cells has been implicated in cancer development and progression [73, 74], and it is also considered a hallmark of cancer. Some of the important characteristic features that are seen during tumor inflammation, such as stimulation of angiogenesis and epithelial-mesenchymal transition [75] are in parallel with the observation of increased HA synthesis and catabolism seen in tumor development and progression. Thus, it appears that cancer cells reprogram their behavioral program in a way that is similar to the embryonic development and tissue repair process.

2.2. Catabolism of HA Polymer to HA Oligomers

HA catabolism is an important step during tumor progression, involving the conversion of HA polymer to HA oligomers by hyaluronidase enzymes

(HYAL). In humans, the three HYAL genes such as HYAL 1, 2 and 3 code for the Hyal-1, -2 and -3 enzymes respectively (for review, see [76]).

The outcome of the HA catabolic process is the varying size of HA oligosaccharides, with a wide variety of biological functions. In the context of tumor development, HA oligosaccharides can have both tumor suppressor and promoter functions (for review, see [77]). The catabolism of HA during tumor progression is depicted in figure 2.

HA catabolism in the body is known to occur via three different pathways such as through the HA binding proteins such as CD44 [78, 79] and RHAMM [80-82], endocytosis process and by the action of free radicals. With respect to the involvement of HA binding protein, once the high molecular HA binds to HABP, the catabolism of HA initiated by the action of Hayal-2. The Hayal-2 cleaves 2×10^4 kDa HA to 20 kDa size HA fragments [83]. These small HA fragments from the plasma membrane are delivered to early endosomes and eventually processed in lysosomes.

In the lysosomes, lysosomal Hayal-1 acts upon these HA fragments and cleaves them to generate mostly tetrasaccharides [84]. In the second pathway, HA receptors such as the HA receptor for endocytosis (HARE) [43] or stabilin 2 [44], the lymphatic vessel endothelial HA receptor (LYVE) 1 [12], and layilin [85] are involved in HA catabolism. With the help of these receptors, the HA from tissue ECM is cleared into the blood vessels and lymphatics, and finally metabolized in the liver, kidney, and possibly spleen.

It is known that eighty-five percent of total HA from the body is cleared by lymph node pathways [86]. Another pathway, which is beyond the scope of this chapter, is by the action of free radicals under oxidative condition [87, 88].

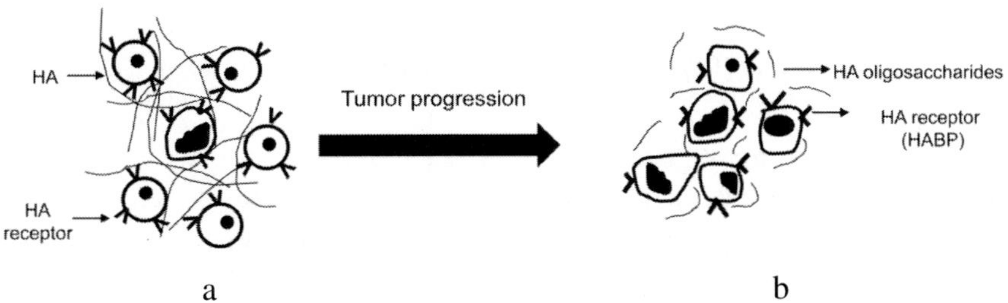

Figure 2. Biological synthesis and catabolism of HA during carcinogenesis. A) In the majority of the malignant tumors, aberrant HA accumulation is known to promote the tumor growth. B) As the tumor progress, the excess HA is degraded to HA oligosaccharides by hyaluronidase enzymes that are produced by the tumor cells. These HA oligosaccharides are shown to have diverse functions during tumor progression.

HA catabolism is an important step during tumor progression. The outcome of HA catabolism is the varying size of HA oligosaccharides, with a wide variety of biological functions. In the context of tumor development, HA oligosaccharides can have both a tumor suppressor and promoter functions. For instance, HA fragments with a 30- to 50-saccharide length are seen in highly invasive bladder cancer as a result of HYAL activity, and known to induce angiogenesis [89]. The HA fragments with a 6-saccharide unit range are known to promote tumorigenicity in human pancreatic cancer cell line [90]. In cases of fibrosarcoma, low molecular weight HA (LMWHA) rather than high molecular weight HA enhanced the adhesion of fibrosarcoma cells by interacting with RHAMM, thus promoting tumor cell migration and invasion [91]. The HA fragments with 6 saccharides thought to promote tumor progression in breast and stomach cancer may be through the HA oligo binding protein [92] (Prasad Kolapalli. S etal. unpublished results). It has been observed that 6.5 kDa oligosaccharide increased the invasion ability of breast cancer cells by inducing matrix metalloproteinase expression and increasing the CD44 cleavage [93]. On the other hand, HA fragments that are 6 to 24 saccharides inhibited cell proliferation in B16F10 melanoma cells [94].

Therefore, as mentioned above, HA oligosaccharides, depending on the size and perhaps the context of the tumor, can either promote or inhibit tumor progression. Therefore, it would be interesting to understand the mechanism by which HA oligosaccharides regulate cancer cell behavior and the factors that are responsible for producing varying size of HA oligosaccharides during tumor progression.

3. TARGETING OF ABERRANT HA SYNTHESIS AND ACCUMULATION IN MALIGNANT TUMOR

Given the extensive role of HA during tumor development and progression, studies are conducted to target HA accumulation in the tumor by many different ways as shown in figure 3. Since HAS gene aberrant expression has been shown in various types of malignant tumors, blocking the HAS gene expression by siRNA could prevent HA accumulation. In this regard, it has been shown that transfection of cells with HAS 2 antisense nucleotides blocked the expression of HAS and HA synthesis [95] in osteosarcoma cells [96]. Alternatively, accumulation of HA in the tumor is also targeted by using recombinant hyaluronidase enzyme (PH20) to degrade HA [97, 98].

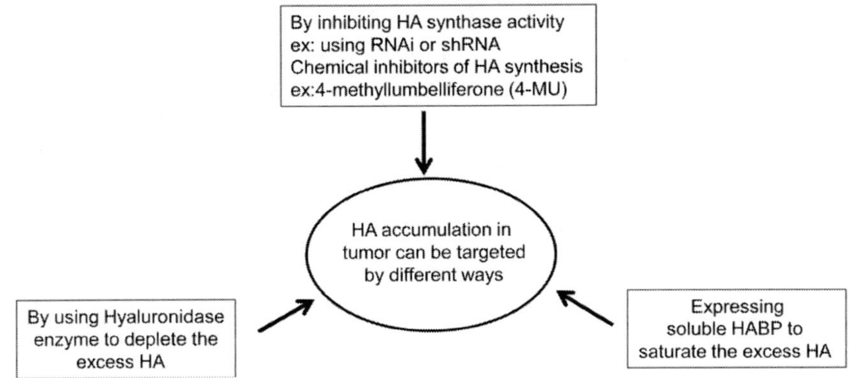

Figure 3. HA accumulation in a malignant tumor can be targeted by many different ways. Some of the methods have been tested successfully in pre-clinical models (see text for detail).

However, the concern with this idea was the stability in *in vivo* and immunologic response because the enzyme was from animal sources. To address this concern recently, a pegylated form of recombinant human PH20 called PEGPH20 hyaluronidase was developed with increased stability in *in vivo* (up to 10.3 h), and it showed high antitumor activity in non-small cell lung cancer patients explants that are grown in nude mice [99]. On the other hand, small molecular inhibitors such as 4-methyllumbelliferone (4-MU) have been shown to block HA synthesis, inhibiting melanoma invasion, metastasis, and breast cancer invasion [100-103]. In addition, it has been shown that 4-MU decreases prostate tumor growth, and in a xenograft model it reduced microvessel density [104].

Recently, 4-MU has been shown to reduce tumorigenicity of osteosarcoma cells by suppressing tumor cell associated matrix and HA accumulation [105].

On the other hand, it has been shown that HA oligosaccharides also have anti-tumor activity against melanoma cells in an animal model [94], and it was demonstrated that this inhibitory activity operates by suppressing the PI3 kinase/Akt cell survival pathway [106].

In another study, HA oligo with an 8-saccharide unit was able to inhibit cell proliferation, migration, and invasive ability of osteosarcoma cells by preventing the interaction between endogenous HA and HABP [107].

After decades of research, now it is well known that the effect of HA on malignant tumors is mediated through interaction with HABPs. Among HABPs, CD44 and RHAMM have been extensively studied for their roles in tumor development and progression [79, 108-114]. Recently, we have reported that another HABP using H11B2C2 mAb is consistently overexpressed in various types of human malignant tumor tissues, compared to benign tumors [47]. It is known that when CD44 interacts with HA, it triggers signaling pathways to promote tumor progression and metastasis [108, 115].

Therefore, attempts were made to block HA and CD44 interaction using different reagents. With respect to this strategy, the overexpression of soluble CD44 was shown to competes with cell surface CD44 for binding with HA [106, 116-118]. In another attempt, antibody against CD44 has been used to target HA binding sties, specifically anti-CD44V6 was exploited for cancer therapy but with limited success [119, 120]. The correlation between CD44V6 protein expression and the aggressiveness of many cancers was the reason to use CD44V6 antibody in cancer therapy [109, 115]. On the other hand, siRNAs are used to inhibit CD44 expression in several studies [116, 121, 122].

Since the siRNAs are known to have stability issues, new varieties of cell-specific shRNAs, effective against CD44V6, were developed [123]. Using this approach, CD44V6 expression was successfully inhibited and shown to reduce cancer growth in an animal model [123].

CONCLUSION

The ECM within tissue is a highly dynamic niche, not only acting as a reservoir for growth factors, but also controlling cell behavior. HA is one of the vital components of ECM. The biological synthesis of HA and its degradation in the malignant tumor is fine-tuned by the HA synthase gene expression and action of hyaluronidase enzymes. A growing body of scientific literature has clearly revealed the role of aberrant synthesis and catabolism of HA during tumor development and progression. Therefore, targeting HA accumulation may be promising in cancer therapy, and there will be a continued effort to achieve this by various means.

ACKNOWLEDGMENTS

We are thankful to all the graduate students for their contributions in our projects. We are grateful to doctors and pathologists for collecting tumor tissue samples and analyzing them. Author R.K.B is grateful to Prof. Karunakumar

Mortha, Department of Biochemistry, University of Mysore, India. We apologize to researchers whose work may have been omitted in the chapter.

REFERENCES

[1] Frantz, C., Stewart, K. M. and Weaver, V. M. (2010). The extracellular matrix at a glance. *J. Cell Sci.* 123, 4195-4200.

[2] Alberts, B., Johnson, A., Lewis, J., Raff, M., Roberts, K., and Walter, P. (2007). *Molecular Biology of the Cell* (London: Garland Science).

[3] Iozzo, R. V. and Murdoch, A. D. (1996). Proteoglycans of the extracellular environment: clues from the gene and protein side offer novel perspectives in molecular diversity and function. *Faseb J.* 10, 598-614.

[4] Schaefer, L. and Schaefer, R. M. (2010). Proteoglycans: from structural compounds to signaling molecules. *Cell Tissue Res.* 339, 237-246.

[5] Esko, J. D., K. K., Lindahl, U. (2009). *Proteoglycans and Sulfated Glycosaminoglycans,* 2nd Edition (Cold Spring Harbor (NY): Cold Spring Harbor Laboratory Press).

[6] Egeblad, M., Rasch, M. G. and Weaver, V. M. (2010). Dynamic interplay between the collagen scaffold and tumor evolution. *Curr. Opin. Cell. Biol.* 22, 697-706.

[7] Lutolf, M. P. and Hubbell, J. A. (2005). Synthetic biomaterials as instructive extracellular microenvironments for morphogenesis in tissue engineering. *Nat. Biotechnol.* 23, 47-55.

[8] Dutta, R. C. and Dutta, A. K. (2009). Cell-interactive 3D-scaffold; advances and applications. *Biotechnol. Adv.* 27, 334-339.

[9] Uygun, B. E., Yarmush, M. L. and Uygun, K. (2012). Application of whole-organ tissue engineering in hepatology. *Nat. Rev. Gastroenterol. Hepatol.* 9, 738-744.

[10] Benders, K. E., van Weeren, P. R., Badylak, S. F., Saris, D. B., Dhert, W. J., and Malda, J. (2013). Extracellular matrix scaffolds for cartilage and bone regeneration. *Trends Biotechnol.* 31, 169-176.

[11] Meyer, K. and Palmer, J. (1934). The polysaccharide of the vitreous humor. *J. Biol. Chem.* 107, 629-634.

[12] Brimacombe, J. S. and Webber, J. M. (1964). *Mucopolysaccharides* (Amsterdam: Elsevier).

[13] Scott, J. E. (1989). Secondary structures in hyaluronan solutions: chemical and biological implications. *The Biology of Hyaluronan* (Ciba Foundation Symposium 143).

[14] Laurent, T. C. and Fraser, J. R. (1992). Hyaluronan. *Faseb J.* 6, 2397-2404.

[15] Cowman, M. K. and Matsuoka, S. (2005). Experimental approaches to hyaluronan structure. *Carbohydr. Res.* 340, 791-809.

[16] Necas, J., B. L., Brauner, P., Kolar, J. (2008). Hyaluronic acid (hyaluronan): a review. *Veterinarni Medicina* 53, 397-411.

[17] Lapcik, L. J. a. L., Lapcik, L., De Smedt, S., Demeester, J., and Chabrecek, P. (1998). Hyaluronan: Preparation, Structure, Properties, and Applications. *Chem. Rev.* 98, 2663-2684.

[18] Weigel, P. H., Hascall, V. C. and Tammi, M. (1997). Hyaluronan synthases. *J. Biol. Chem.* 272, 13997-14000.

[19] Spicer, A. P. and McDonald, J. A. (1998). Characterization and molecular evolution of a vertebrate hyaluronan synthase gene family. *J. Biol. Chem.* 273, 1923-1932.

[20] Matsumoto, K., Li, Y., Jakuba, C., Sugiyama, Y., Sayo, T., Okuno, M., Dealy, C. N., Toole, B. P., Takeda, J., Yamaguchi, Y., and Kosher, R. A. (2009). Conditional inactivation of Has2 reveals a crucial role for hyaluronan in skeletal growth, patterning, chondrocyte maturation and joint formation in the developing limb. *Development* 136, 2825-2835.

[21] Krause, R. (1972). *Streptococci and Streptococcal Disease* (New York: Academic Press).

[22] Csoka, A. B. and Stern, R. (2013). Hypotheses on the evolution of hyaluronan: a highly ironic acid. *Glycobiology* 23, 398-411.

[23] Fraser, J. R., Laurent, T. C. and Laurent, U. B. (1997). Hyaluronan: its nature, distribution, functions and turnover. *J. Intern. Med.* 242, 27-33.

[24] Bremer, F. M. and Rasquin, F. (1998). Histochemical localization of hyaluronic acid in vitreous during embryonic development. *Invest. Ophthalmol. Vis. Sci.* 39, 2466-2469.

[25] Toole, B. P. (2001). Hyaluronan in morphogenesis. *Seminars in Cell and Developmental Biology* 12, 79-87.

[26] Banerjee, S. D. and Toole, B. P. (1991). Monoclonal antibody to chick embryo hyaluronan-binding protein: changes in distribution of binding protein during early brain development. *Dev. Biol.* 146, 186-197.

[27] Ausprunk, D. H., Boudreau, C. L. and Nelson, D. A. (1981). Proteoglycans in the microvascular. II. Histochemical localization in proliferating capillaries of the rabbit cornea. *Am. J. Pathol.* 103, 367-375.

[28] Liu, C. M., Chang, C. H., Yu, C. H., Hsu, C. C., and Huang, L. L. (2009). Hyaluronan substratum induces multidrug resistance in human

mesenchymal stem cells via CD44 signaling. *Cell Tissue Res*. 336, 465-475.

[29] Preston, M. and Sherman, L. S. (2011). Neural stem cell niches: roles for the hyaluronan-based extracellular matrix. *Front Biosci.* (Schol. Ed.) 3, 1165-1179.

[30] Laurent, T. C. and Fraser, J. R. E. (1986). The properties and turnover of hyaluronan. In: *Functions of Proteoglycans*, Volume 124, pp. 9-29 (Chichester, England: Wiley).

[31] Brecht, M., Mayer, U., Schlosser, E., and Prehm, P. (1986). Increased hyaluronate synthesis is required for fibroblast detachment and mitosis. *Biochem. J.* 239, 445-450.

[32] Yoneda, M., Shimizu, S., Nishi, Y., Yamagata, M., Suzuki, S., and Kimata, K. (1988). Hyaluronic acid-dependent change in the extracellular matrix of mouse dermal fibroblasts that is conducive to cell proliferation. *J. Cell Sci.* 90 (Pt 2), 275-286.

[33] Furukawa, K. and Terayama, H. (1979). Pattern of glycosaminoglycans and glycoproteins associated with nuclei of regenerating liver of rat. *Biochim. Biophys. Acta* 585, 575-588.

[34] Ripellino, J. A., Bailo, M., Margolis, R. U., and Margolis, R. K. (1988). Light and electron microscopic studies on the localization of hyaluronic acid in developing rat cerebellum. *J. Cell. Biol.* 106, 845-855.

[35] Eggli, P. S. and Graber, W. (1995). Association of hyaluronan with rat vascular endothelial and smooth muscle cells. *J. Histochem. Cytochem.* 43, 689-697.

[36] Evanko, S. P. and Wight, T. N. (1999). Intracellular localization of hyaluronan in proliferating cells. *J. Histochem. Cytochem.* 47, 1331-1342.

[37] Hascall, V. C., Majors, A. K., De La Motte, C. A., Evanko, S. P., Wang, A., Drazba, J. A., Strong, S. A., and Wight, T. N. (2004). Intracellular hyaluronan: a new frontier for inflammation? *Biochim. Biophys. Acta* 1673, 3-12.

[38] Pavasant, P., Shizari, T. and Underhill, C. B. (1996). Hyaluronan contributes to the enlargement of hypertrophic lacunae in the growth plate. *J. Cell Sci.* 109 (Pt 2), 327-334.

[39] Huang, L., Grammatikakis, N., Yoneda, M., Banerjee, S. D., and Toole, B. P. (2000). Molecular characterization of a novel intracellular hyaluronanbinding protein. *J. Biol. Chem.* 275, 29829-29839.

[40] Aruffo, A., Stamenkovic, I., Melnick, M., Underhill, C. B., and Seed, B. (1990). CD44 is the principal cell surface receptor for hyaluronate. *Cell* 61, 1303-1313.

[41] Hardwick, C., Hoare, K., Owens, R., Hohn, H. P., Hook, M., Moore, D., Cripps, V., Austen, L., Nance, D. M., and Turley, E. A. (1992). Molecular cloning of a novel hyaluronan receptor that mediates tumor cell motility. *J. Cell Biol.* 117, 1343-1350.

[42] Banerji, S., Ni, J., Wang, S. X., Clasper, S., Su, J., Tammi, R., Jones, M., and Jackson, D. G. (1999). LYVE-1, a new homologue of the CD44 glycoprotein, is a lymph-specific receptor for hyaluronan. *J. Cell Biol.* 144, 789-801.

[43] Zhou, B., Weigel, J. A., Fauss, L., and Weigel, P. H. (2000). Identification of the hyaluronan receptor for endocytosis (HARE). *J. Biol. Chem.* 275, 37733-37741.

[44] Politz, O., Gratchev, A., McCourt, P. A., Schledzewski, K., Guillot, P., Johansson, S., Svineng, G., Franke, P., Kannicht, C., Kzhyshkowska, J., Longati, P., Velten, F. W., Johansson, S., and Goerdt, S. (2002). Stabilin-1 and -2 constitute a novel family of fasciclin-like hyaluronan receptor homologues. *Biochem. J.* 362, 155-164.

[45] Falkowski, M., Schledzewski, K., Hansen, B., and Goerdt, S. (2003). Expression of stabilin-2, a novel fasciclin-like hyaluronan receptor protein, in murine sinusoidal endothelia, avascular tissues, and at solid/ liquid interfaces. *Histochem. Cell Biol.* 120, 361-369.

[46] Sloane, J. A., Blitz, D., Margolin, Z., and Vartanian, T. (2010). A clear and present danger: endogenous ligands of Toll-like receptors. *Neuromolecular Med.* 12, 149-163.

[47] Boregowda, R. K., Appaiah, H. N., Karunakumar, M., Parameshwariah, S., Avadani, G., Sunila, S., and Banerjee, S. (2013). Development and validation of H11B2C2 monoclonal antibody-reactive hyaluronic acid binding protein: overexpression of HABP during human tumor progression. *Tumour Biol.* 34., 597-608.

[48] Grammatikakis, N., Grammatikakis, A., Yoneda, M., Yu, Q., Banerjee, S. D., and Toole, B. P. (1995). A novel glycosaminoglycan-binding protein is the vertebrate homologue of the cell cycle control protein, Cdc 37. *J. Biol. Chem.* 270, 16198-16205.

[49] Deb, T. B. and Datta, K. (1996). Molecular cloning of human fibroblast hyaluronic acid-binding protein confirms its identity with P-32, a protein co-purified with splicing factor SF2. Hyaluronic acid-binding protein as

P-32 protein, co-purified with splicing factor SF2. *J. Biol. Chem.* 271, 2206-2212.

[50] Entwistle, J., Hall, C. L. and Turley, E. A. (1996). HA receptors: regulators of signalling to the cytoskeleton. *J. Cell Biochem.* 61, 569-577.

[51] Assmann, V., Marshall, J. F., Fieber, C., Hofmann, M., and Hart, I. R. (1998). The human hyaluronan receptor RHAMM is expressed as an intracellular protein in breast cancer cells. *J. Cell. Sci.* 111 (Pt 12), 1685-1694.

[52] Bissell, M. J. and Radisky, D. (2001). Putting tumours in context. *Nat. Rev. Cancer* 1, 46-54.

[53] Knudson, W., Biswas, C., Li, X. Q., Nemec, R. E., and Toole, B. P. (1989). The role and regulation of tumour-associated hyaluronan. *Ciba Found. Symp.* 143, 150-159; discussion 159-169, 281-155.

[54] Toole, B. P. (2002). Hyaluronan promotes the malignant phenotype. *Glycobiology* 12, 37R-42R.

[55] Borcgowda, R. K., Appaiah, H. N., Siddaiah, M., Kumarswamy, S. B., Sunila, S., Kn, T., Mortha, K., Toole, B., and Banerjee, S. D. (2006). Expression of Hyaluronan in human tumor progression. *J. Carcinog.* 5, 2.

[56] Siiskonen, H., Poukka, M., Tyynela-Korhonen, K., Sironen, R., and Pasonen-Seppanen, S. (2013). Inverse expression of hyaluronidase 2 and hyaluronan synthases 1-3 is associated with reduced hyaluronan content in malignant cutaneous melanoma. *BMC Cancer* 13, 181.

[57] Bertrand, P., Girard, N., Delpech, B., Duval, C., d'Anjou, J., and Dauce, J. P. (1992). Hyaluronan (hyaluronic acid) and hyaluronectin in the extracellular matrix of human breast carcinomas: comparison between invasive and non-invasive areas. *Int. J. Cancer* 52, 1-6.

[58] Ponting, J., Kumar, S. and Pye, D. (1993). Colocalization of hyaluronan and hyaluronectin in normal and neoplastic breast tissues. *Int. J. Oncol.* 2, 889-893.

[59] Tammi, R. H., Kultti, A., Kosma, V. M., Pirinen, R., Auvinen, P., and Tammi, M. I. (2008). Hyaluronan in human tumors: pathobiological and prognostic messages from cell-associated and stromal hyaluronan. *Semin. Cancer Biol.* 18, 288-295.

[60] Llaneza, A., Vizoso, F., Rodriguez, J. C., Raigoso, P., Garcia-Muniz, J. L., Allende, M. T., and Garcia-Moran, M. (2000). Hyaluronic acid as prognostic marker in resectable colorectal cancer. *Br. J. Surg.* 87, 1690-1696.

[61] Vizoso, F. J., del Casar, J. M., Corte, M. D., Garcia, I., Corte, M. G., Alvarez, A., and Garcia-Muniz, J. L. (2004). Significance of cytosolic hyaluronan levels in gastric cancer. *Eur. J. Surg. Oncol.* 30, 318-324.

[62] Corte, M. D., Gonzalez, L. O., Junquera, S., Bongera, M., Allende, M. T., and Vizoso, F. J. (2010). Analysis of the expression of hyaluronan in intraductal and invasive carcinomas of the breast. *J. Cancer Res. Clin. Oncol.* 136, 745-750.

[63] Itano, N., Sawai, T., Atsumi, F., Miyaishi, O., Taniguchi, S., Kannagi, R., Hamaguchi, M., and Kimata, K. (2004). Selective expression and functional characteristics of three mammalian hyaluronan synthases in oncogenic malignant transformation. *J. Biol. Chem.* 279, 18679-18687.

[64] Li, Y. and Heldin, P. (2001). Hyaluronan production increases the malignant properties of mesothelioma cells. *Br. J. Cancer* 85, 600-607.

[65] Liu, N., Gao, F., Han, Z., Xu, X., Underhill, C. B., and Zhang, L. (2001). Hyaluronan synthase 3 overexpression promotes the growth of TSU prostate cancer cells. *Cancer Res.* 61, 5207-5214.

[66] Ichikawa, T., Itano, N., Sawai, T., Kimata, K., Koganehira, Y., Saida, T., and Taniguchi, S. (1999). Increased synthesis of hyaluronate enhances motility of human melanoma cells. *J. Invest. Dermatol.* 113, 935-939.

[67] Enegd, B., King, J. A., Stylli, S., Paradiso, L., Kaye, A. H., and Novak, U. (2002). Overexpression of hyaluronan synthase-2 reduces the tumorigenic potential of glioma cells lacking hyaluronidase activity. *Neurosurgery* 50, 1311-1318.

[68] Kosaki, R., Watanabe, K. and Yamaguchi, Y. (1999). Overproduction of hyaluronan by expression of the hyaluronan synthase Has2 enhances anchorage-independent growth and tumorigenicity. *Cancer Res.* 59, 1141-1145.

[69] Itano, N., Atsumi, F., Sawai, T., Yamada, Y., Miyaishi, O., Senga, T., Hamaguchi, M., and Kimata, K. (2002). Abnormal accumulation of hyaluronan matrix diminishes contact inhibition of cell growth and promotes cell migration. *Proc. Natl. Acad. Sci. US* 99, 3609-3614.

[70] Virchow, R., Virchow, R (1863). *Aetiologie der neoplastichen Geschwulste/Pathogenie der neoplastischen Geshwulste.* Veriag von August Hisrchwald, Berlin, Germany.

[71] Dunham, L. J. (1972). Cancer in man at site of prior benign lesion of skin or mucous membrane: a review. *Cancer Res.* 32, 1359-1374.

[72] Dvorak, H. F. (1986). Tumors: wounds that do not heal. Similarities between tumor stroma generation and wound healing. *N Engl. J. Med.* 315, 1650-1659.

[73] De Visser, K. E., Eichten, A. and Coussens, L. M. (2006). Paradoxical roles of the immune system during cancer development. *Nat. Rev. Cancer* 6, 24-37.

[74] Balkwill, F., Charles, K. A. and Mantovani, A. (2005). Smoldering and polarized inflammation in the initiation and promotion of malignant disease. *Cancer Cell* 7, 211-217.

[75] Schafer, M. and Werner, S. (2008). Cancer as an overhealing wound: an old hypothesis revisited. *Nat. Rev. Mol. Cell Biol.* 9, 628-638.

[76] Stern, R. (2003). Devising a pathway for hyaluronan catabolism: are we there yet? *Glycobiology* 13, 105R-115R.

[77] Lokeshwar, V. B. and Selzer, M. G. (2008). Hyalurondiase: both a tumor promoter and suppressor. *Semin. Cancer Biol.* 18, 281-287.

[78] Lesley, J., Hascall, V. C., Tammi, M., and Hyman, R. (2000). Hyaluronan binding by cell surface CD44. *J. Biol. Chem.* 275, 26967-26975.

[79] Ponta, H., Sherman, L. and Herrlich, P. A. (2003). CD44: from adhesion molecules to signalling regulators. *Nat. Rev. Mol. Cell Biol.* 4, 33-45.

[80] Zhang, S., Chang, M. C., Zylka, D., Turley, S., Harrison, R., and Turley, E. A. (1998). The hyaluronan receptor RHAMM regulates extracellular-regulated kinase. *J. Biol. Chem.* 273, 11342-11348.

[81] Cheung, W. F., Cruz, T. F. and Turley, E. A. (1999). Receptor for hyaluronan-mediated motility (RHAMM), a hyaladherin that regulates cell responses to growth factors. *Biochem. Soc. Trans.* 27, 135-142.

[82] Lynn, B. D., Li, X., Cattini, P. A., Turley, E. A., and Nagy, J. I. (2001). Identification of sequence, protein isoforms, and distribution of the hyaluronan-binding protein RHAMM in adult and developing rat brain. *J. Comp. Neurol.* 439, 315-330.

[83] Lepperdinger, G., Strobl, B. and Kreil, G. (1998). HYAL2, a human gene expressed in many cells, encodes a lysosomal hyaluronidase with a novel type of specificity. *J. Biol. Chem.* 273, 22466-22470.

[84] Stern, R., Asari, A. A. and Sugahara, K. N. (2006). Hyaluronan fragments: An information-rich system. *European Journal of Cell Biology* 85, 699-715.

[85] Bono, P., Cordero, E., Johnson, K., Borowsky, M., Ramesh, V., Jacks, T., and Hynes, R. O. (2005). Layilin, a cell surface hyaluronan receptor, interacts with merlin and radixin. *Exp. Cell Res.* 308, 177-187.

[86] Weigel, P. H. and Yik, J. H. (2002). Glycans as endocytosis signals: the cases of the asialoglycoprotein and hyaluronan/chondroitin sulfate receptors. *Biochim. Biophys. Acta* 1572, 341-363.

[87] Myint, P., Deeble, D. J., Beaumont, P. C., Blake, S. M., and Phillips, G. O. (1987). The reactivity of various free radicals with hyaluronic acid: steady-state and pulse radiolysis studies. *Biochim. Biophys. Acta* 925, 194-202.

[88] Deguine, V., Menasche, M., Ferrari, P., Fraisse, L., Pouliquen, Y., and Robert, L. (1998). Free radical depolymerization of hyaluronan by Maillard reaction products: role in liquefaction of aging vitreous. *Int. J. Biol. Macromol.* 22, 17-22.

[89] Lokeshwar, V. B., Obek, C., Soloway, M. S., and Block, N. L. (1997). Tumor-associated hyaluronic acid: a new sensitive and specific urine marker for bladder cancer. *Cancer Res.* 57, 773-777.

[90] Sugahara, K. N., Hirata, T., Hayasaka, H., Stern, R., Murai, T., and Miyasaka, M. (2006). Tumor cells enhance their own CD44 cleavage and motility by generating hyaluronan fragments. *J. Biol. Chem.* 281, 5861-5868.

[91] Kouvidi, K., Berdiaki, A., Nikitovic, D., Katonis, P., Afratis, N., Hascall, V. C., Karamanos, N. K., and Tzanakakis, G. N. (2011). Role of receptor for hyaluronic acid-mediated motility (RHAMM) in low molecular weight hyaluronan (LMWHA)-mediated fibrosarcoma cell adhesion. *J. Biol. Chem.* 286, 38509-38520.

[92] Srinivas, P., Kollapalli, S. P., Thomas, A., Mortha, K. K., and Banerjee, S. D. (2012). Bioactive hyaluronan fragment (hexasaccharide) detects specific hexa-binding proteins in human breast and stomach cancer: possible role in tumorogenesis. *Indian J. Biochem. Biophys.* 49, 228-235.

[93] Kung, C. I., Chen, C. Y., Yang, C. C., Lin, C. Y., Chen, T. H., and Wang, H. S. (2012). Enhanced membrane-type 1 matrix metalloproteinase expression by hyaluronan oligosaccharides in breast cancer cells facilitates CD44 cleavage and tumor cell migration. *Oncol. Rep.* 28, 1808-1814.

[94] Zeng, C., Toole, B. P., Kinney, S. D., Kuo, J. W., and Stamenkovic, I. (1998). Inhibition of tumor growth in vivo by hyaluronan oligomers. *Int. J. Cancer* 77, 396-401.

[95] Nishida, Y., Knudson, C. B., Nietfeld, J. J., Margulis, A., and Knudson, W. (1999). Antisense inhibition of hyaluronan synthase-2 in human articular chondrocytes inhibits proteoglycan retention and matrix assembly. *J. Biol. Chem.* 274, 21893-21899.

[96] Nishida, Y., Knudson, W., Knudson, C. B., and Ishiguro, N. (2005). Antisense inhibition of hyaluronan synthase-2 in human osteosarcoma

cells inhibits hyaluronan retention and tumorigenicity. *Exp. Cell Res.* 307, 194-203.

[97] Li, L., Asteriou, T., Bernert, B., Heldin, C. H., and Heldin, P. (2007). Growth factor regulation of hyaluronan synthesis and degradation in human dermal fibroblasts: importance of hyaluronan for the mitogenic response of PDGF-BB. *Biochem. J.* 404, 327-336.

[98] Shuster, S., Frost, G. I., Csoka, A. B., Formby, B., and Stern, R. (2002). Hyaluronidase reduces human breast cancer xenografts in SCID mice. *Int. J. Cancer* 102, 192-197.

[99] Thompson, C. B., Shepard, H. M., O'Connor, P. M., Kadhim, S., Jiang, P., Osgood, R. J., Bookbinder, L. H., Li, X., Sugarman, B. J., Connor, R. J., Nadjsombati, S., and Frost, G. I. (2010). Enzymatic depletion of tumor hyaluronan induces antitumor responses in preclinical animal models. *Mol. Cancer Ther.* 9, 3052-3064.

[100] Kudo, D., Kon, A., Yoshihara, S., Kakizaki, I., Sasaki, M., Endo, M., and Takagaki, K. (2004). Effect of a hyaluronan synthase suppressor, 4-methylumbellifcronc, on B16F-10 melanoma cell adhesion and locomotion. *Biochem. Biophys. Res. Commun.* 321, 783-787.

[101] Urakawa, H., Nishida, Y., Wasa, J., Arai, E., Zhuo, L., Kimata, K., Kozawa, E., Futamura, N., and Ishiguro, N. (2012). Inhibition of hyaluronan synthesis in breast cancer cells by 4-methylumbelliferone suppresses tumorigenicity in vitro and metastatic lesions of bone in vivo. *Int. J. Cancer* 130, 454-466.

[102] Kultti, A., Pasonen-Seppanen, S., Jauhiainen, M., Rilla, K. J., Karna, R., Pyoria, E., Tammi, R. H., and Tammi, M. I. (2009). 4-Methylumbelliferone inhibits hyaluronan synthesis by depletion of cellular UDP-glucuronic acid and downregulation of hyaluronan synthase 2 and 3. *Exp. Cell Res.* 315, 1914-1923.

[103] Yoshihara, S., Kon, A., Kudo, D., Nakazawa, H., Kakizaki, I., Sasaki, M., Endo, M., and Takagaki, K. (2005). A hyaluronan synthase suppressor, 4-methylumbelliferone, inhibits liver metastasis of melanoma cells. *FEBS Lett.* 579, 2722-2726.

[104] Lokeshwar, V. B., Lopez, L. E., Munoz, D., Chi, A., Shirodkar, S. P., Lokeshwar, S. D., Escudero, D. O., Dhir, N., and Altman, N. (2010). Antitumor activity of hyaluronic acid synthesis inhibitor 4-methylumbelliferone in prostate cancer cells. *Cancer Res.* 70, 2613-2623.

[105] Arai, E., Nishida, Y., Wasa, J., Urakawa, H., Zhuo, L., Kimata, K., Kozawa, E., Futamura, N., and Ishiguro, N. (2011). Inhibition of hyalu-

ronan retention by 4-methylumbelliferone suppresses osteosarcoma cells in vitro and lung metastasis in vivo. *Br. J. Cancer* 105, 1839-1849.

[106] Ghatak, S., Misra, S. and Toole, B. P. (2002). Hyaluronan oligosaccharides inhibit anchorage-independent growth of tumor cells by suppperssing the phosphoinositide 3-kinase/Akt cell survival pathway. *J. Biol. Chem.* 277, 38013-38020.

[107] Hosono, K., Nishida, Y., Knudson, W., Knudson, C. B., Naruse, T., Suzuki, Y., and Ishiguro, N. (2007). Hyaluronan oligosaccharides inhibit tumorigenicity of osteosarcoma cell lines MG-63 and LM-8 in vitro and in vivo via perturbation of hyaluronan-rich pericellular matrix of the cells. *Am. J. Pathol.* 171, 274-286.

[108] Toole, B. P. (2009). Hyaluronan-CD44 Interactions in Cancer: Paradoxes and Possibilities. *Clin. Cancer Res.* 15, 7462-7468.

[109] Orian-Rousseau, V. (2010). CD44, a therapeutic target for metastasising tumours. *Eur. J. Cancer* 46, 1271-1277.

[110] Marhaba, R. and Zoller, M. (2004). CD44 in cancer progression: adhesion, migration and growth regulation. *J. Mol. Histol.* 35, 211-231.

[111] Hall, C. L., Yang, B., Yang, X., Zhang, S., Turley, M., Samuel, S., Lange, L. A., Wang, C., Curpen, G. D., Savani, R. C., Greenberg, A. H., and Turley, E. A. (1995). Overexpression of the hyaluronan receptor RHAMM is transforming and is also required for H-ras transformation. *Cell* 82, 19-26.

[112] Wang, C., Thor, A. D., Moore, D. H., 2nd, Zhao, Y., Kerschmann, R., Stern, R., Watson, P. H., and Turley, E. A. (1998). The overexpression of RHAMM, a hyaluronan-binding protein that regulates ras signaling, correlates with overexpression of mitogen-activated protein kinase and is a significant parameter in breast cancer progression. *Clin. Cancer Res.* 4, 567-576.

[113] Maxwell, C. A., Rasmussen, E., Zhan, F., Keats, J. J., Adamia, S., Strachan, E., Crainie, M., Walker, R., Belch, A. R., Pilarski, L. M., Barlogie, B., Shaughnessy, J., Jr., and Reiman, T. (2004). RHAMM expression and isoform balance predict aggressive disease and poor survival in multiple myeloma. *Blood* 104, 1151-1158.

[114] Gurski, L. A., Xu, X., Labrada, L. N., Nguyen, N. T., Xiao, L., van Golen, K. L., Jia, X., and Farach-Carson, M. C. (2012). Hyaluronan (HA) interacting proteins RHAMM and hyaluronidase impact prostate cancer cell behavior and invadopodia formation in 3D HA-based hydrogels. *PLoS One* 7, e50075.

[115] Misra, S., Heldin, P., Hascall, V. C., Karamanos, N. K., Skandalis, S. S., Markwald, R. R., and Ghatak, S. (2011). Hyaluronan-CD44 interactions as potential targets for cancer therapy. *Febs J.* 278, 1429-1443.

[116] Ghatak, S., Misra, S. and Toole, B. P. (2005). Hyaluronan constitutively regulates ErbB2 phosphorylation and signaling complex formation in carcinoma cells. *J. Biol. Chem.* 280, 8875-8883.

[117] Peterson, R. M., Yu, Q., Stamenkovic, I., and Toole, B. P. (2000). Perturbation of hyaluronan interactions by soluble CD44 inhibits growth of murine mammary carcinoma cells in ascites. *Am. J. Pathol.* 156, 2159-2167.

[118] Ahrens, T., Sleeman, J. P., Schempp, C. M., Howells, N., Hofmann, M., Ponta, H., Herrlich, P., and Simon, J. C. (2001). Soluble CD44 inhibits melanoma tumor growth by blocking cell surface CD44 binding to hyaluronic acid. *Oncogene* 20, 3399-3408.

[119] Tijink, B. M., Buter, J., de Bree, R., Giaccone, G., Lang, M. S., Staab, A., Leemans, C. R., and van Dongen, G. A. (2006). A phase I dose escalation study with anti-CD44v6 bivatuzumab mertansine in patients with incurable squamous cell carcinoma of the head and neck or esophagus. *Clin. Cancer Res.* 12, 6064-6072.

[120] Orian-Rousseau, V. and Ponta, H. (2008). Adhesion proteins meet receptors: a common theme? *Adv. Cancer Res.* 101, 63-92.

[121] Ghatak, S., Hascall, V. C., Markwald, R. R., and Misra, S. (2010). Stromal hyaluronan interaction with epithelial CD44 variants promotes prostate cancer invasiveness by augmenting expression and function of hepatocyte growth factor and androgen receptor. *J. Biol. Chem.* 285, 19821-19832.

[122] Misra, S., Ghatak, S. and Toole, B. P. (2005). Regulation of MDR1 expression and drug resistance by a positive feedback loop involving hyaluronan, phosphoinositide 3-kinase, and ErbB2. *J. Biol. Chem.* 280, 20310-20315.

[123] Misra, S., Hascall, V. C., De Giovanni, C., Markwald, R. R., and Ghatak, S. (2009). Delivery of CD44 shRNA/nanoparticles within cancer cells: perturbation of hyaluronan/CD44v6 interactions and reduction in adenoma growth in Apc Min/+ MICE. *J. Biol. Chem.* 284, 12432-12446.

Chapter 4

HYALURONIDASES: BIOLOGICAL FEATURES AND CURRENT THERAPEUTIC APPLICATIONS

R. Gazzola, G. Colombo, S. Marcelli and L. Vaienti*
Plastic Surgery Department
I.R.C.C.S. Policlinico San Donato – Università degli Studi di Milano
Piazza Malan, San Donato Milanese, Italy

ABSTRACT

Hyaluronidases are enzymes that are able to degrade hyaluronic acid. They are classified into three groups: mammalian, leech and microbial hyaluronidases. Their role in vivo is essential for the homeostasis and metabolism of the extracellular matrix, thus regulating cell growth.

Their action can be measured by chemical, physicochemical and biological methods. The first employs reductimetric reactions on the products of hyaluronidases. Physicochemical procedures measure the changes of physical features of the substrate, such asviscosity. Biological methods evaluate the spreading effect of the enzyme in the animal model.

Several applications have been described in the literature, mainly as a spreading factor. The most important application is hypodermoclysis, the capacity of increasing absorption and dispersion of an injected drug. Hyaluronidases are used in radiography for increasing the spread of

---

* Email address: riccardogazzola@gmail.com

radiopaque substances, in gynecology along with ergometrine to prevent post-partum haemorrhage, in obstretric blocks of the pudendal and ileoinguinal nerve.

In ophtalmology, hyaluronidases are instilled with local anesthetics for retrobulbar, peribulbar, sub-Tenon's, and van Lint blocks. In fact, injecting hyaluronidase leads to smaller increases in intraocular pressure (IOP), results in less distortion of the surgical site, decreased incidence of postoperative strabismus, and has the potential for limiting local anesthetic myotoxicity.

Hyaluronidase can also be used for post-operative edema reduction. Hyaluronidases are thus employed in transplant surgery to reduce interstitial edema and the risk of rejection.

In chemotherapy hyaluronidases prevent the risk of local injury after drug extravasation. In pain therapy the enzymes are employed to increase the spread of local anesthetics in selective blocks, while in cardiology they can reduce myocardial infarction size after coronary occlusion.

Moreover hyaluronidases can be successfully used to treat complications arising from inappropriate subcutaneous injection of hyaluronic acid for aesthetic purposes. In fact an excessive inoculation of hyaluronic acid filler may result invisible overcorrection, nodules, bumps or ischemic complications due to compression of the dermal plexus.

Given the importance of these enzymes for homeostasis and their applications in medicine, a sound knowledge is essential for everyday practice.

INTRODUCTION

Hyaluronidases are a class of enzymes that are able to degrade hyaluronic acid (HA). These enzymes are not specific for HA however. They could also degrade, at a slower rate, chondroitin (CS) and chondroitin sulfates (ChS) [1].

Hyaluronidases are synthesized in nature by mammals, leeches /hookworms and microbes and maybe divided accordingly into 3 classes. Each class of enzymes has a different mechanism of action.

Mammalian Hyaluronidases

Mammalian hyaluronidases are endo-β-N-acetlyhexosaminidases that degrade the β-1,4 glycosidic linkages of HA, resulting in tetrasaccharides. These enzymes can be found in the spermatozoa and lysosomes of several mammalians, and in the venom of reptiles, snakes and hymenoptera. They are synthetized by six gene sequences in the human genome, called HYAL genes. Each enzyme is tissue- specific (e.g. the sperm adhesion molecule, SPAM 1) [2, 3].

These forms are:

- Hyal 1, which was the first human somatic hyaluronidase enzyme isolated in human plasma and exists as two isoenzymes, a high weight isoform (found in blood) and a low weight isoform(found in the blood circulation and urine) [4].
- Hyal 2, which in humans constitutes the functional receptor binding surface proteins of the Jaagsiekte sheep retrovirus (JRSV) and of the enzootic nasal tumor virus (ENTV) [5].
- Hyal 3 is widely expressed in chondrocytes, testis, bone marrow, usually when fibroblasts differentiate into chondrocytes [6]
- Hyal 4 is probably a GPI-anchored protein like PH-20 and Hyal 2 [7]
- Hyal-P1this sequence is a pseudogene, in that it is not translated into a protein.
- PH-20 or SPAM 1, also known as testis hyaluronidase. This form comes with a GPI- anchor and is located on the inner surface of the acrosome. Its function consists in penetrating inside the cumulus ECM and zona pellucida of the ovum [8]. Some hyases are produced by testicular cells and in lower quantities are secreted by the female genital tract [9]. In fact, PH-20 has also been found with PCR assays in the epididymis, seminal vescicles, prostate, female genital tract, breast, placenta and fetal tissue and in some malignancies [10].

Hyal 2- and 3 undergo up-regulation under inflammatory cytokines (IL-1, TNF-alfa) [11]. All these hyaluronidases are active at acidic pH except for PH-20 and Hyal 2 [12].

Endo-β-D-Glucuronidases

These enzymes target the β-1,3 glycosidic bond, producing tetra- and hexasaccharides. In contrast to mammalian glycosidases, this group is HA-specific and exerts no activity on other GAGs. These enzymes are found in the salivary glands of leeches, hookworms, snakes, bees, stonefish, scorpions, spiders, lizards, wasps, caterpillars and hornets. The enzyme acts as a spreading factor to facilitate the diffusion of a toxin through the systemic circulation due to the degradation of HA (especially in the ECM surrounding blood vessels) [13].

A typical example is that of the brown spider (Loxosceles genus), endemic in the South of Brasil. Its bite produces swelling, erythema, hemorrhage and a dermonecrotic lesion with gravitational spreading. Hyaluronidases play a central role in the action of the venom, acting as a "spreading factor" facilitating the penetration of the venom into the surrounding tissues. The target of these enzymes is the glycosaminoglycan found in the extracellular matrix of the victim [14]

In some cases (scorpions, bees, hornets and wasps), hyaluronidases can induce severe IgE-mediated fatal systemic anaphylactic reactions in humans [15].

Microbial Hyaluronidases

Microbial hyaluronidases are hyaluronatelyases. They act by β-elimination at β-1,4 glycosidic linkages and differ from the other hyaluronidases in that they do not use hydrolysis. This results in unsaturated disaccharides. Several microorganisms synthesize these enzymes, such as strains of *Clostridium, Micrococcus, Streptococcus*, and *Streptomyces*. These enzymes are involved in pathogenesis for gram-positive bacteria, and play a less important role in pathogenesis for gram-negative bacteria.

Microbial hyaluronidases are virulence factors that cause disease progression by degrading the ECM components. In other cases these enzymes facilitate adhesion and provide nutrients [16]. Hyases may also hide the bacterial surface from host defenses. They play a central role in pneumonia, meningitis and sepsis [17].

Mammalian hyaluronidases are more similar to microbial hyaluronidases. Their target is the extracellular matrix (ECM), which contains HA. This molecule is a straight chain glycosaminoglycan, made of disaccharide of D-glucuronic acid and N-acetyl-D-glucosamine. ECM undergoes a continuous turnover, produced by multiple enzymes, mainly constituted by the hyase family members (E.C.3.2.1.35) [18].

Hyaluronidases may be also classified, according to their pH-dependent activity, into acid active (between pH 3 and 4) and neutral active (pH 5-8). Human liver and serum hyaluronidases belong to the first group, while snake and bee venom hyaluronidases belong to the second [3, 19,20].

As previously discussed, hyaluronidases are involved in several mechanisms, such as diffusion of substances, fertilization and cell growth.

Hyaluronan is considered to trap the extravascular fluid and prevent it from being drained by the lymphatics. The injection of hyaluronidases has been demonstrated to facilitate the resorption of the extravascular fluids and thus the resolution of edema.rHuPH20 sustained release gel (containing human recombinant hyaluronidase) has been demonstrated to reduce post-operative edema by increasing interstitial fluid drainage [21]. Hyaluronidase has a role in the development of osteoarthritis (OA), a chronic disability at older age that affects 30% of the population above the age of 65. Although several biologic and mechanical factors lead to OA, elevated levels of hydrolytic enzyme activity may degrade the articular cartilage and worsen the disease [22]. Among degradative endoglucosaminidase, hyaluronidaseplaysa central role during extracellular matrix degradation associated with the synovial joint disease [23]. Bark extract of Payenadasyphylla (100 µg/ml) showed the highest inhibitory activity against bovine testicular hyaluronidase, also inhibiting the expression of HYAL1, HYAL2, MMP-3, MMP-13 mRNA protein expression. The inhibition of hyaluronidase may thus have a role in the therapy of OA.

Hyaluronidase, Cell Growth and Cancer Progression

Hyaluronidases are involved in cell growth and cancer progression. Although high levels of HA are correlated with several neoplasms, the role of hyaluronidase in cancer progression is not a clear-cut unequivocal one. In fact it could alternatively increase and counteract cancer invasiveness by several mechanisms. For example, Hyal 1 and Hyal 2 have been demonstrated to be up-regulated in tobacco-related cancers and in respiratory tract cancers, probably induced by tumor suppressor genes [24, 25], although high levels of these enzymes have been found in other cancers (e.g. bladder and prostate).

Cancer-Promoting Issues

Several mechanisms mediated by hyaluronidases could theoretically increase cancer growth and invasiveness. Sometimes their expression correlates with tumor growth. E.g. PH-20 is expressed by malignant tumors (breast, prostate, melanoma), working as an oncogene or tumor suppressor gene product.

Their transcription and synthesis is nonetheless finely modulated. In fact the production of HA by the tumor changes with the progression of the disease. Initially high-weight hyaluronic acid is synthesized to facilitate diffusion of nutrients for cancer growth. When diffusion becomes insufficient for cancer survival, lower-weight HA is produced due to the modulated expression of hyaluronidase, which may promote angiogenesis [26] and cancer progression.

Hyaluronidases endowed with a GPI anchor may also be involved in signaling, in that the aggregation and cross-linking of these domains produce intracellular events [8], although their relationship to cancer growth is yet to be clarified.

HA has been demonstrated to suppress TGF-beta 1 expression. Therefore the action of hyaluronidase in degrading HA may increase the invasiveness of several cancers [27].

Cancer-Inhibiting Issues

Conversely, hyaluronidase is able to inhibit cancer growth and metastasis. This is mainly related to the degradation of hyaluronate rather than to a direct role played by the enzyme in inhibiting tumor progression.

Cancer cells are able to use anaerobic metabolism to generate lactate, a phenomenon known as the Warburg effect [28]. This leads to increased extracellular lactates that stimulate fibroblasts to synthetize hyaluronic acid [29]. HA is directly proportional to tumor aggressiveness, and enhances cell growth. Over-production of hyaluronidase allows the accumulation of extracellular HA, essential for carcinogenesis [30]. Moreover HA could encourage metastases, increasing cell motility and growth (by absorbing nutrients from the connective tissue). The degradation of HA could therefore exert negative effects on tumor development [31]. Moreover, hyaluronidase has been demonstrated to antagonize cancer growth by increasing tumor necrosis factor toxicity (TNF) [7].

HA protects cancer cells from non-specific suppressive and anti-inflammatory factors, masking tumor antigens and making tumors less accessible to T cells. This is supported by the demonstration of greater cytotoxic action against tumor cells after treatment with hyaluronidase [32].

HA influences cell migration and polarization. Tumor cells express CD44 receptors on their surface, binding and anchoring the cell to HA molecules. This process has been demonstrated to promote growth and metastasis [33]. Therefore hyaluronidase can theoretically slow cancer progression.

In some cases neoplasms may produce inhibitors of hyaluronidase, important for cancer progression. The inhibition of these enzymes could theoretically produce the accumulation of intermediate-sized HA fragments that stimulate angiogenesis and inflammation, essential for tumor growth [34].

In vivo hyalruonidase enhances the anticancer effects of adriamycin [35], delays the appearance of carcinogen-induced tumors [36], and prevents lymphnode invasion in the murine model [37].

Metabolism and Methods of Measure

High-molecular-weight hyaluronan is synthesized by healthy cells at the plasma membrane and degraded by plasma membrane-hyaluronidases, usually in peripheral tissues and lymph. When these enzymes enter the circulatory system they are degraded by the sinusoidal liver endothelium. A decoupled synthesis of hyaluronan and its catabolites produced by hyaluronidase is observed in some malignancies, as fragmented forms of hyaluronan [37].

Hyaluronidase inhibitors are essential for extracellular matrix homeostasis and HA metabolism. The hyaluronidase inhibitors are carefully regulated, in a similar manner to metalloproteinases (MMPs) and tissue inhibitors (TIMPs). The activity of hylauronidase inhibitors is usually magnesium-dependent [37].

Hyaluronidases have been widely used in medicine for several purposes. The main uses are listed below, although a detailed discussion will follow in this chapter:

- Spreading agent to promote the diffusion of several substances injected subcutaneously [38]. These enzymes are able to increase the absorption of several substances, promoting drug diffusion into the extracellular matrix and into the circulatory system by increasing the permeability of vessels. This feature is employed both for facilitating the penetration of drugs in tissue and to avoid tissue damage by accidental extravasation of drugs [3]. It is also employed to reduce tissue edema [39] or to correct excessive injection of hyaluronic-acid-containing fillers [40].

- Disease indicator. In fact HA levels raise during wound healing and rapid cell turnover [41]. Levels can also increase in the case of inflammation or stress such as blood loss, shock, and septicemia [42]. High levels are observed in liver cirrhosis, liver fibrosis.
- Removal of the cumulus-corona-oocyte complex formed during intracytoplasmic sperm injection [43]
- Ocular surgery, to maintain the operative space and to protect the endothelial layer of the cornea. In particular HA is used to treat vitreous hemorrhage [44] and in the implantation of artificial intraocular lens.

Several methods have been described to determine the activity of HA. El-Saforya [7] grouped these methods into chemical, physicochemical and biological methods.

Chemical Methods

The breakage of hyaluronan molecules at the glucosidic bond results in reducing sugars that can be detected with reductimetric procedures. Alternatively a colorimetric assay can be used to measure the N-acetylglucosamine (NAG) end-groups from hyaluronan. Known as the Morgan-Elson reaction, it may be employed only for microbial and mammalian hyaluronidases that generate NAG end-groups. The method consists in measuring a red product obtained after reaction with p-dimethylaminobenzaldehyde. Less common assessment methods are spectrophotometry, chromatography, electrophoresis and fluo-rogenichyaluronate reactions.

Physicochemical Methods

These methods are usually older and less precise than the above-mentioned methods [7]. They employ the physical modifications induced by hyaluronidase on the substrate: e.g. precipitates, spinnability, and viscosity. The main methods of measurement are the viscosity reduction method and the turbidimetric method.

The first measures the decrease of viscosity in a solution containing HA and hyaluronidase. This method is highly dependent on the viscosity of the initial hyaluronate solution. In fact the molecular mass of the HA employed influences the viscosity, making it difficult to obtain a standard measurement.

The mucin clot method uses the clotting of HA in acid solution with protein. Hyaluronidase reduces clotting and changes the precipitate features until no precipitate is observed. Instead the spinnability method measures the maximum length of filaments of bovine synovial fluid that are drawn at standard velocity. Hyaluronidases reduce the length of filaments proportionally to their concentration.

Biological assays measure the spreading of a indicator dye injected along with hyaluronidase in the subcutaneous tissue of animals. The spreading capacity is proportional to the concentration of hyaluronidase [7].

Applications of Hyaluronidase in Medicine

Hyaluronidaseuse is subject to a strict protocol. A skin test is recommended before the first dose to test for possible sensitivity or allergy to enzymes; a thorough patient interview is necessary to determine any previous enzyme administration and possible sensitization. The composition of the hyaluronidase must be visually checked for particles or discoloration to decide whether it is safe for use. Informed consent must be obtained before proceeding within filtration, particularly in off-label applications.

The quantity of hyaluronidase to be injected should be adjusted to the employment of the enzymes; for instance, in the case of previously HA-injected areas the practitioner should adjust the quantity to the type of hyaluronic acid and the number and extent of affected areas.

The injection of more than 200 units of hyaluronidase per treatment should in all circumstances be avoided. Infiltration does not usually require dilution of the enzymes with local anesthetic.

The infiltration technique varies depending on different employment of enzymes.

In the case of subcutaneous infiltration, areas to be injected should be marked on the overlying skin, and infiltration by hyaluronidase should be extremely accurate and limited to the exact recipient area. The recipient area should be investigated by ultrasonography in case of previous local infiltration of fillers (e.g. hyaluronic acid) in order to assess depth, quantity and extension of the filler itself. A 30-gauge needle, ranging from 3 to 13 mm in length, may be effectively employed. In our clinical experience, for all subcutaneous use, a few units of hyaluronidase (10-20 units) are adequate. The injection should be made perpendicular to the skin; multiple treatments may be necessary depending on application.

A particular subcutaneous infiltration is required in the case of hypodermoclysis. An ordinary two-way subcutaneous infusion set is required with the fluid container supported about 1 meter above the patient. Suitable hypodermic needles are attached to the connecting rubber tubes after any contained air has been expelled. The needles are then inserted into the subcutaneous tissues at selected sites and the fluid allowed to run in by gravity. When this method is employed, approximately 200mL of fluid can be administered in 20 minutes.

The rate of administration can be further varied by raising or lowering the reservoir. However, whenever greater speed is required, with possibly smaller volumes, the rate may be increased by injecting the fluid from a 20 or 50 mL syringe, having drawn the appropriate amount of a solution of hyaluronidase into the syringe beforehand. Repeated infusions may be given as required but it is advisable to change the injection site.

As solutions of the enzyme lose their activity in 12-24 hours, it is important that fresh solutions be prepared prior to each administration. The possible sites of administration are numerous, irrespective of the position of the patient. One may, for example, choose from the axillae pectoral regions, subscapular and trunk areas, thighs and calves. In children, Gaisford and Evans consider the anterior abdominal wall presents the best site both from the point of view of maximal absorption and freedom from napkin contamination [45, 46].

Hypodermoclysis

The most important application of hyaluronidase is hypodermoclysis, i.e., the capacity of increasing absorption and dispersion of the injected drug [45].

The extracellular matrix presents a significant barrier to the effective subcutaneous and intramuscular delivery of many drugs, limiting both pharmacokinetic parameters and injection volumes. The space outside the adipocytes in the hypodermis is not a fluid, but rather a solid extracellular matrix of collagenous fibrils embedded within a glycosaminoglycan-rich viscoelastic gel which buffers convective forces. The extracellular matrix limits the volume of drug that can be injected at a single site, as well as the rate and amount that reach the vascular compartment. Hyaluronidase acts as a spreading factor by reversibly modifying the hypodermis leading to increased injection volumes. Greater bioavailability from subcutaneous and intramuscular injection may overcome some key limitations of these routes of administration in multiple care settings [44].

In particular a number of studies have investigated the use of hyaluronidases as a spreading agent along with chemotherapeutics, local anesthetics, contrast media and in the management of injuries caused by intravenous extravasation of antibiotics, aminophylline, mannitol, parental nutrition solution, electrolyte infusions and contrast medium [47].

A randomized clinical trial was conducted of recombinant human hyaluronidase-facilitated subcutaneous versus intravenous rehydration in mild to moderately dehydrated children in the emergency department. [48].

Radiology

In pyelography, a mixture of radiopaque substances with hyaluronidase is injected subcutaneously and bilaterally into the scapular region. The area of injection is massaged vigorously to aid dispersion. Normally renal secretion commences five minutes after injection. Adequate concentration is obtained in 24-25 minutes and is equal to that obtained by the intravenous method.

Hyaluronidases are employed successfully for the management of contrast medium extravasation (CMEV), a well-known complication of contrast-enhanced computed tomography (CT) scanning. CMEV can also occur in magnetic resonance imaging (MRI), but the complications are rare given the low volumes used. CT contrast extravasation occurs relatively infrequently, in ~0.5% (range 0.13-0.68%) of cases, but can have severe side effects. Nonionic low-osmolar contrast media is known to reduce the risk of severe soft tissue injury but the risk of soft tissue injury is often related to the volume of CMEV.

In minor cases, contrast medium extravasation may cause pain, swelling, and localized erythema. However, in more severe cases, extensive tissue and skin necrosis, ulceration, and compartment syndrome may occur, often necessitating surgery.

Hyaluronidase has been used successfully in the management of extravasated contrast media according to the literature [49-51].

In addition, recombinant human hyaluronidaseuse is approved as an adjunct in subcutaneous urography for improving resorption of radiopaque agents [52].

Hyaluronidase can be successfully used to treat extravasation of a large volume of iodinated contrast media and appears to be a reasonable treatment option for more extensive subcutaneous contrast media extravasations; it must be noted however that iodinated contrast media reduces the activity of hyaluronidases [53].

Gynecology

In the case of obstetric anesthesia, where hyaluronidase is used in a mixture with local anesthetics, the needle is passed horizontally to the ischial spine, and 5mL of the solution is deposited there to anaesthetize the pudendal nerve as it enters Alcock's canal. Next, 5mL of the mixture is infiltrated in the superior portion of the labia minora to anaesthetize the perineal branches of the ileoinguinal nerve. The same procedure is carried out on both sides of the perineum.

In the prevention of post-partum hemorrhage, hyaluronidase is instead administered intramuscularly, together with ergometrine, at the crowning of the fetal head [54]. It has been demonstrated that the oxytocic effect of the drug is apparent 3-4 minutes earlier than when ergometrine is given alone, and approximates to the effect seen after intravenous ergometrine [55].

The solution is prepared a little before it is needed, and at the moment of the crowning or delivery of the fetal head it is injected intramuscularly into the thigh.

A pilot study has also investigated the use of a perineal injection of hyaluronidase to prevent the occurrence of spontaneous perineum lacerations, especially of second- and third-degree lacerations during delivery. There were no cases higher than first degree lacerations occurring within the group of patients who received the hyaluronidase, but further studies have shown that the use of injectable hyase did not increase the proportion of intact perineum and did not reduce the proportion of severe perineal trauma [56-59].

In reproductive medicine, recombinant hyaluronidase is currently used for intracytoplasmic sperm injection to remove the cumulus-corona-oocyte complex, composed of granulose cells inside a matrix of oligosaccharide chains cross-linked by hyaluronan binding proteins and proteoglycans [43].

Ophthalmology

The enzyme hyaluronidase can also be instilled directly into the anterior chamber of the eye, together with local anesthetics in the formulation of ophthalmic collyrium, to obtain safe and long lasting anesthesia of the eye in routine surgical procedures such as cataract correction.

Hyaluronidase is most frequently used in combination with anesthetics for ophthalmologic surgery (eg, retrobulbar block, peribulbar block, sub-Tenon's block, and van Lint block). Rationale for the inclusion of hyaluronidase in combination with local anesthesia techniques includes smaller increases in intraocular pressure (IOP), less distortion of the surgical site, decreased incidence of postoperative strabismus, and potential for limiting local anesthetic myotoxicity because of quicker spread. In some studies, inclusion of hyaluronidase increases globe and lid akinesia, which may improve the safety of the procedure.

Ovine hyaluronidase also has been investigated and approved for the treatment of vitreous hemorrhages [60, 61].

Retrobulbar Block

Although retrobulbar blocks are less common than peribulbar blocks due to potential anatomic risk factors associated with administration, and the need for an additional facial nerve block to prevent blinking, they may still be preferred for certain procedures that would benefit from lower volumes of local anesthetic (ie, 3-5 mL).

Nine randomized, prospective, controlled studies evaluating hyaluronidase added to retrobulbar blocks in more than 1,300 patients were reviewed.

A variety of anesthetic mixtures and hyaluronidase doses (ranging from 0.75 to 200 IU/mL) were used. The end points included akinesia, induction time, need for supplementary block, and volume of local anesthetic. Addition of hyaluronidase to retrobulbar blocks generally resulted in improved akinesia and was well tolerated. Fewer complications, such as a lower tendency for prolapse, were observed in patients who received hyaluronidase. Overall, a dose between 3.75 and 75 IU/mL hyaluronidase in retrobulbar blocks appears sufficient to provide a beneficial effect on akinesia.

Van Lint Blocks

Because retrobulbar blocks do not provide lid akinesia, they are often combined with van Lint blocks. This method of facial nerve block was the first to be reported and is considered to be the classic technique. Typically, 5 to 10 mL of anesthetic (often the same mixture as used for the retrobulbar block) is used, although some suggest that volumes as low as 2 mL are sufficient.

Peribulbar Block

The peribulbar block technique was developed to minimize the risk of injury to structures within the intraconal space. It is performed by injection into the extraconal space using larger volumes of local anesthetic (e.g., up to 12 mL). The larger volume is necessary for its spread into the entire corpus adiposum of the orbit and eyelids to block the orbicularis muscle.

Fourteen randomized, prospective, controlled studies evaluating hyaluronidase in peribulbar blocks in nearly 1,800 patients were reviewed and analyzed.

These studies involved a variety of local anesthetic combinations and hyaluronidase doses (ranging from 3.75 to 300 IU/mL). The end points included akinesia, induction time, need for supplementary block, and volume of local anesthetic. Evidence of improvement in peribulbar block with hyaluronidase is equivocal. Approximately half of the studies show a benefit when including hyaluronidase, whereas the rest of the studies demonstrate equivalence across a range of hyaluronidase doses. Hyaluronidase was well tolerated. In some of these studies, fewer complications such as increased IOP were observed in patients who received hyaluronidase.

Although hyaluronidase does not consistently improve peribulbar block efficacy, it is beneficial for its effects on facilitating the spread of larger volumes of anesthetic and reducing complications.

Sub-Tenon's Block

A sub-Tenon's block (STB) administered using either a needle or cannula provides high-quality anesthesia of the whole globe using relatively small volumes (e.g. 1.25-5 mL). The addition of hyaluronidase in STB has been studied in varying combinations across six randomized, prospective, controlled studies encompassing more than 550 patients.

Testing the addition of hyaluronidase to STB has been conducted using assorted local anesthesia combinations and doses of hyaluronidase (ranging from 15-150 IU/mL). End points included akinesia, induction time, volume of local anesthetic, and quality of the block. Hyaluronidase addition to STB generally resulted in improved akinesia and was well tolerated. Although a wide range of hyaluronidase doses were used, and despite a lack of dose-response data, the studies demonstrate that doses between 15 and 150 IU/mL can provide benefit when used in STB.

Hyaluronidase thus plays a clear role in ophthalmology and has been demonstrated to enhance quality, efficacy, and safety of local anesthetic blocks [62].

Overall, several studies demonstrate improved akinesia and block onset time when adding hyaluronidase to retrobulbar, peribulbar, or sub-Tenon's blocks, along with fewer complications or adverse effects. In addition, hyaluronidase allows for lower volumes of anesthetic to be used, a fact which helps to minimize elevations in IOP during surgical procedures.

The literature discusses the use of hyaluronidase in local anesthetic blocks and demonstrates a high degree of variability in methods and dosages. However, this variation itself suggests that hyaluronidase can be used safely and successfully.

Given the value of hyaluronidase in ophthalmology, it is important to note that the source of hyaluronidase merits careful consideration.

Animal-derived hyaluronidase preparations (specifically compounded formulations) are associated with hypersensitivity reactions or toxicities. In addition they have significantly lower purity and potency than does human recombinant hyaluronidase. It is a matter of concern that with compounded animal-derived hyaluronidase there is a potential lack of information regarding impurities, along with unknown compliance with quality-assurance measures.

Human recombinant hyaluronidase rHuPH20, however, was shown to be highly pure and highly potent (approximately 140- to 200-fold increase compared with compounded animal-derived hyaluronidase; approximately 5.6-fold increase compared with manufactured animal-derived hyaluronidase). It has a favorable safety profile and is associated with no reports of serious hypersensitivity reactions. It is therefore preferable to animal-derived enzymes in everyday practice [63].

Edema Reduction

Off-label applications of hyaluronidases include edema reduction in the case of paraphimosis, intestinal intussusceptions and supra-glottal airway edema, in addition to edema reduction in transplant organ rejection [44].

Transplant Surgery

Interstitial edema of rejecting organs can be correlated with the accumulation of hyaluronan in the transplant since hyaluronan has a strong water-binding capacity. Treatment with the hyaluronan-degrading enzyme hyaluronidase has been proved to reduce not only the hyaluronan content but also the water content of the graft. In particular, hyaluronidase has been used in edema treatment during cardiac allograft rejection in rats, and further studies have proved that it is also possible to retain the effect of the enzyme in heparinized animals provided that sufficient doses of the enzyme are administered [64-66]

Chemotherapy

Extravasation of cytotoxic agents during peripheral intravenous administration may cause severe local injuries (e.g. tissue necrosis and ulceration). Extravasation can be prevented in most cases with the systematic use of careful administration techniques. However, the management of this complication remains an important challenge in oncology. Many antidotes have been evaluated experimentally and only a few are able to reduce the local toxicity of most vesicant cytotoxic drugs. Because no randomized trials on the management of cytotoxic drug extravasation in humans have ever been completed, recommendations must be based on the most consistent experimental evidence and on cumulative clinical experience from available case reports and uncontrolled studies, reviewed here.

In case of inadvertent extravasation of Vinca alkaloids during iv administration, hyaluronidase can reduce the risk of progression to skin necrosis [67].

In chemotherapy, hyaluronidases are also injected along with the chemotherapeutic agent thus improving the penetration of the drug inside the malignant tissue (e.g. brain tumors) [38].

Pain Therapy

In pain therapy, the pioneering experience dates back to 1949 when Atkinson discovered that the addition of hyaluronidase to local anesthetics such as lidocaine or procaine allowed a safer and more durable neural block [53]. Nowadays hyaluronidase is employed as a spreading agent along with local anesthetics in caudal blocks, directly in the painful region (e.g., joints, tendons) or via an intrathecal route [39]. It has been proved that the combination of the hyaluronidase with local anesthetics does not affect the activity of the enzyme. However the enhancement of absorption induced by hyaluronidases may lead to a higher incidence of systemic allergic reactions to local anesthetics. An off-label application includes the lysis of epidural adhesions to treat chronic painful conditions such as radiating pain and lower back pain.

Cardiology

In cardiology, a considerable effort has been devoted to reducing myocardial injury and infarct size after ischemia. To this end, hyaluronidase has been pharmacologically studied. Some benefits due to hyaluronidase treatment have been demonstrated in myocardic ischemia using animal models [68-72]. It has been suggested that the mechanisms underlying these results are related to the biological effect of the metabolic products of hyaluronidase, which enhance blood flow and decrease tissue edema [71-76]. Hyaluronidase exerts a relevant influence upon the lymphatic system of the myocardium and has been proved to have a cardioprotective effect in the treatment of acute myocardial infarction, where it reduces the size of myocardial infarcts after coronary occlusion. Improving the blood flow leads to an early removal of toxic metabolites. In addition, it promotes nutrient and oxygen transportation to the injured ischemic myocardium [68, 73]. Trials on human and primates do not show any substantial and objective benefit in reducing infarct size or improving myocardial function [77-81], although the combination of hyaluronidase and urokinase decreases the mortality rate after myocardial infarction in an animal model [78, 82, 83]. In the early 1990s, hyaluronidase experiments on myocardial infarction tailed off, and more effective drugs and coronary angioplasty/stenting were developed, yielding excellent results [84].

Plastic Surgery and Dermatology

In plastic surgery and dermatology, hyaluronidases are employed for a wide range of treatments.

Combined use with local anesthetic, as previously discussed, is one of the most common indications together with HA filling correction.

A quite uncommon, although rationaluse, is that of reducing distortion caused by local anesthetics in septorhinoplasty [85].

The most important application of hyaluronidase for the dermatologist and plastic surgeon is probably the revision of HA filler injection. In such cases the indication can be strictly esthetic, functional or both.

Hyaluronidases can be successfully employed esthetically to eliminate nodules or bumps, to treat an overcorrection by HA filler injection and to treat excessive superficial infiltration.

Methods of injection and the hyaluronidase effective dose may vary depending on the anatomical region and on the amount and characteristics of the HA employed for filling.

Painful and inflammatory nodules should be treated primarily with antibiotics (e.g. clarithromycin 500mg or minocycline 100mg twice per day covers a wide range of early infections). The duration of antibiotic treatment ranges from two to six weeks depending on the severity of the infection; association of hyaluronidase injection permits an early dissolution of the HA filler with a prompt recovery [86]. When a large amount of HA filler has been inoculated, incision and drainage of the lesion maybe indicated. Concerning therapeutic dosage of hyaluronidase, in these cases a range from 3 to 75 i.u. is effective [87].

In the case of inflammatory long-lasting nodules and when a particulate filler is injected, steroid intralesional inoculation should be attempted, as a final procedure, before excision; steroids should be used if the patient is already receiving an antibiotic therapy [86-88].

Over-injection of HA filler may cause an excessive and unwanted augmentation. Generally, non-painful lumps after injection are likely to disappear within the first two weeks after injection; the physician should recommend massage of the area, evaluate the patient regularly and address any fears the patient may have. The therapeutic dosage of hyaluronidase should be adjusted according to the area and volume of injected HA. Even low doses (< 3 i.u.) may be effective in reversing excessive augmentations (e.g. in the lower lid) [86, 87].

Hyaluronidase may be employed for complications deriving from HA filler treatments.

Intra-arterial injection and excessive filling are quite uncommon events that result in an ischemic injury of cutaneous tissue due to embolization or compression of the sub-dermal plexus; the glabella and nasal ala are the most susceptible areas [89, 90].

In these patients timeliness is crucial. The skin becomes discolored (pale and/or purple) evolving to necrosis in a few hours (rarely over 24h) and pain is usually referred. Signs and symptoms may vary depending on the patient and type of vascular disorder (arterial or venous) [89-91].

Hyaluronidase represents an efficient drug if used within the first 4-6 hours after HA filler injection; over this time, administration of the enzymes does not seem to be effective in avoiding skin necrosis [90, 92]. However delayed treatments may reduce tissue size damage and improve the healing process.

At the first signs of cutaneous discolouring, occurring immediately after treatment or in the very first hours, a topical application of 2% nitroglycerine paste and warm compresses is recommended, along with hyaluronidases [89, 91]. Antibiotic therapy (topical parenteral or both) must be prescribed in cases of skin breakdown [91].

Drug Interactions

Hyaluronidases are a group of enzymes synthesized by a variety of organisms. Their function is variable and heterogeneous in nature [7]. These enzymes are employed in medicine with several pharmacological indications [40]. However, knowledge of pharmacokinetic mechanisms is still limited. Clinical trials demonstrate that hyaluronidase is rapidly eliminated from plasma after i.v. administration with a half-time plasma clearance of 2.1 +/- 0.2 min; the short plasma half-life originates, apart from urine and bile excretion, from tissue uptake (maximum in the liver). Most hyaluronidases achieve their optimal enzymatic activity at pH 3.7-4.0, being partially active at neutral pH [93]. Due to chemical or physical incompatibilities, hyaluronidase should not be administered with benzodiazepines, furosemide, phenytoin, dopamine or alpha-adrenergic agonists [3].

A thorough preliminary medical history interview should be conducted to investigate assumption of the above-mentioned drugs. Hyaluronidase activity is inhibited by heparin, ascorbic acid, anti inflammatory agents (e.g. salicylates, dexamethasone, indomethacin) antihistamines, mast-cell stabilizers, various plant-based drugs (e.g. antioxidants, flavonoids), dicumarene and radiographic iodinated contrast media (patients taking those drugs necessitate a higher hyaluronidase dose to achieve the same dispersing effect).

The main activator agents are adrenaline, dopamine, histamine and 0,9% NaCl. No interactions have been observed with local anesthetics [3, 93,94].

Complications

Allergic reactions are the only complications associated with the use of hyaluronidases in humans.

Such reactions are unusual, and their clinical manifestation may vary depending on the area treated and type of administration; also the onset time of the response is variable (intraoperative- within a few hours- within a few days) [95].

Allergic reactions, mild and benign in most cases, include local complications (local edema, tumor, mild pain) itching sensation, generalized maculo-papular rash and urticaria [20, 96,97]. Local reactions at the injection site are the most common; nevertheless, anaphylaxis has been reported in a few cases [96, 98,99].

A proportionally high number of reactions to hyaluronidase has been reported in ophthalmic surgery where these enzymes are often employed for retrobulbar anesthesia in association with bupivacaine or adrenaline [20].

Studies of ophthalmic surgery reveal an incidence of allergic reactions ranging from 0.05% to 0.69%; the main symptoms are edema, erythema, pain, and itching [20, 100-102].

The incidence of mild allergic reactions could be higher if considering unrecognized cases [101]. However urticaria and angioedema have been reported in less than 0.1% of cases [44].

A significant rise of allergic events has been observed when increasing hyaluronidase dosages; administrations of 200.000 i.u. correspond to 31.3% in allergic complications [98].

While primary sensitization seems to be a trigger prerequisite for most allergic reactions, the underlying immunological mechanisms are Type I (mediated by IgE) and Type IV (mediated by T-cells) hypersensitivity [98, 102, 103].

The above-mentioned reactions could be potentially worsened by the intrinsic hyaluronidase property of increasing capillary permeability [104].

The allergy may be diagnosed by a routine prick or intradermal test [53, 100], while serum IgE antibodies, highly specific for hyaluronidases, may be observed after the first exposure [96].

Moreover IgE synthesis could be triggered by intravenous co-administration of hyaluronidase, chemotherapeutics and dexamethasone [98].

Due to a certain homology in hyaluronidase molecules between species, a cross-allergenicity risk should be evaluated when administering hyaluronidase in patients with a known allergy to insects sting or bites [105].

Although rare, clinicians should be aware of the risks and potential complications of hyaluronidase administration.

A thorough patient interview should be conducted to investigate any previous therapy undergone with hyaluronidase, along with any known allergies to insect stings or bites.

Technical Notes

The prescription of hyaluronidases in the United States of America is - like any drug - regulated by the Food and Drug Administration (FDA) Center for Drug Evaluation and Research (CDER), which reviews applications for new pharmaceutical drugs. Only when the drug fulfills the safety and effectiveness requirements laid down by the CDER will the dosage, route and indications be established [106].

Several subtypes of hyaluronidases (recombinant or animal derived) have been approved for clinical use (e.g. Hylenex®150 i.u./ml, Vitrase®200 i.u./ml).

These enzymes are currently FDA label-approved for the following indications:

- Adjuvant function, increasing absorption and/or dispersion of other injected drugs;
- Hypodermoclysis;
- Subcutaneous urography for improving reabsorption of radiopaque agents.

Known hypersensitivity to hyaluronodase or any other ingredient within the formulation is a contraindication to the use of the product.

Although the promotion of any drug for indications not approved by the FDA is not permitted, off-label prescription is not formally prohibited but is delegated to the practitioner's choice if considered safe and effective [107].

Jurisdiction within the European Union is slightly different. Marketing approval is released after testing the product's quality, efficacy and safety. Authorization is granted by centralized or national procedures. Analogously to the United States, off-label prescriptions are allowed (under personal responsibility) provided that the patient's autonomy and best interests are respected; however off-label promotion is not allowed by article 87 of Directive 2001/83/EC.

Hyaluronidaseis EC approved:

- As adjuvant therapy in subcutaneous drugs administration;
- To increase penetration of local anesthetic;
- To promote absorption of contrast medium in urology;
- To promote subcutaneous hematoma reabsorption.

CONCLUSION

Hyaluronidases are extremely important for the homeostasis of the extracellular matrix and regulation of cell growth. Recently, increasing attention has focused on this enzyme in several fields of medicine. A sound knowledge of the physiologic role of hyaluronidases and their main applications in medicine is essential for everyday practice.

REFERENCES

[1]	Stern, R. and M.J. Jedrzejas, Hyaluronidases: their genomics, structures, and mechanisms of action. *Chem. Rev.*, 2006. 106(3): p. 818-39.

[2]	Csoka, A.B., G.I. Frost, and R. Stern, The six hyaluronidase-like genes in the human and mouse genomes. *Matrix Biol.*, 2001. 20(8): p. 499-508.

[3]	Girish, K.S. and K. Kemparaju, The magic glue hyaluronan and its eraser hyaluronidase: a biological overview. *Life Sci*, 2007. 80(21): p. 1921-43.

[4]   Csoka, A.B., G.I. Frost, T. Wong, and R. Stern, Purification and microsequencing of hyaluronidase isozymes from human urine. *FEBS Lett.*, 1997. 417(3): p. 307-10.

[5]   Duh, F.M., C. Dirks, M.I. Lerman, and A.D. Miller, Amino acid residues that are important for Hyal2 function as a receptor for jaagsiekte sheep retrovirus. *Retrovirology*, 2005. 2: p. 59.

[6]   Nicoll, S.B., O. Barak, A.B. Csoka, R.S. Bhatnagar, and R. Stern, Hyaluronidases and CD44 undergo differential modulation during chondrogenesis. *Biochem. Biophys. Res. Commun.*, 2002. 292(4): p. 819-25.

[7]   El-Saforya, N.S., A.E. Fazaryb, and C.K. Leea, Hyaluronidases, a group of glycosidases: Current and future perspectives. *Carb. Pol.*, 2010. 81: p. 165-181.

[8]   Cherr, G.N., A.I. Yudin, and J.W. Overstreet, The dual functions of GPI-anchored PH-20: hyaluronidase and intracellular signaling. *Matrix Biol.*, 2001. 20(8): p. 515-25.

[9]   Martin-Deleon, P.A., Germ-cell hyaluronidases: their roles in sperm function. *Int. J. Androl.*, 2011. 34(5 Pt 2): p. e306-18.

[10]   Oettl, M., J. Hoechstetter, I. Asen, G. Bernhardt, and A. Buschauer, Comparative characterization of bovine testicular hyaluronidase and a hyaluronate lyase from Streptococcus agalactiae in pharmaceutical preparations. *Eur. J. Pharm. Sci.*, 2003. 18(3-4): p. 267-77.

[11]   Flannery, C.R., C.B. Little, C.E. Hughes, and B. Caterson, Expression and activity of articular cartilage hyaluronidases. *Biochem. Biophys. Res. Commun.*, 1998. 251(3): p. 824-9.

[12]   Vigdorovich, V., R.K. Strong, and A.D. Miller, Expression and characterization of a soluble, active form of the jaagsiekte sheep retrovirus receptor, Hyal2. *J. Virol.*, 2005. 79(1): p. 79-86.

[13]   Kemparaju, K. and K.S. Girish, Snake venom hyaluronidase: a therapeutic target. *Cell Biochem. Funct.*, 2006. 24(1): p. 7-12.

[14]   Ferrer, V.P., T.L. de Mari, L.H. Gremski, D. Trevisan Silva, R.B. da Silveira, W. Gremski, et al., A novel hyaluronidase from brown spider (Loxosceles intermedia) venom (Dietrich's Hyaluronidase): from cloning to functional characterization. *PLoS. Negl. Trop. Dis.*, 2013. 7(5): p. e2206.

[15]   Kolarich, D., R. Leonard, W. Hemmer, and F. Altmann, The N-glycans of yellow jacket venom hyaluronidases and the protein sequence of its major isoform in Vespula vulgaris. *FEBS J.*, 2005. 272(20): p. 5182-90.

[16] Hynes, W.L., A.R. Dixon, S.L. Walton, and L.J. Aridgides, The extracellular hyaluronidase gene (hylA) of Streptococcus pyogenes. *FEMS Microbiol. Lett.*, 2000. 184(1): p. 109-12.

[17] Makris, G., J.D. Wright, E. Ingham, and K.T. Holland, The hyaluronate lyase of Staphylococcus aureus - a virulence factor? *Microbiology*, 2004. 150(Pt 6): p. 2005-13.

[18] Stern, R., Devising a pathway for hyaluronan catabolism: are we there yet? *Glycobiology*, 2003. 13(12): p. 105R-115R.

[19] Kreil, G., Hyaluronidases - a group of neglected enzymes. *Protein Science*, 1995. 4: p. 1666–1669.

[20] Kempeneers, A., L. Dralands, and J. Ceuppens, Hyaluronidase induced orbital pseudotumor as complication of retrobulbar anesthesia. *Bull. Soc. Belge. Ophtalmol.*, 1992. 243: p. 159-66.

[21] Nekoroski, T., R.D. Paladini, D.N. Sauder, G.I. Frost, and G.A. Keller, A recombinant human hyaluronidase sustained release gel for the treatment of post-surgical edema. *Int. J. Dermatol.*, 2013.

[22] Tortorella, M.D., T.C. Burn, M.A. Pratta, I. Abbaszade, J.M. Hollis, R. Liu, et al., Purification and cloning of aggrecanase-1: a member of the ADAMTS family of proteins. *Science*, 1999. 284(5420): p. 1664-6.

[23] Citalingam, K., S. Zareen, K. Shaari, and S. Ahmad, Effects of Payena dasyphylla (Miq.) on hyaluronidase enzyme activity and metalloproteinases protein expressions in interleukin-1beta stimulated human chondrocytes cells. *BMC Complement Altern. Med.*, 2013. 13: p. 213.

[24] Frost, G.I., G. Mohapatra, T.M. Wong, A.B. Csoka, J.W. Gray, and R. Stern, HYAL1LUCA-1, a candidate tumor suppressor gene on chromosome 3p21.3, is inactivated in head and neck squamous cell carcinomas by aberrant splicing of pre-mRNA. *Oncogene*, 2000. 19(7): p. 870-7.

[25] Chang, N.S., Transforming growth factor-beta1 blocks the enhancement of tumor necrosis factor cytotoxicity by hyaluronidase Hyal-2 in L929 fibroblasts. *BMC Cell Biol.*, 2002. 3: p. 8.

[26] West, D.C., I.N. Hampson, F. Arnold, and S. Kumar, Angiogenesis induced by degradation products of hyaluronic acid. *Science*, 1985. 228(4705): p. 1324-6.

[27] Chang, N.S., Transforming growth factor-beta protection of cancer cells against tumor necrosis factor cytotoxicity is counteracted by hyaluronidase (review). *Int. J. Mol. Med*, 1998. 2(6): p. 653-9.

[28] Warburg, O., On the origin of cancer cells. *Science,* 1956. 123(3191): p. 309-14.

[29] Stern, R., S. Shuster, B.A. Neudecker, and B. Formby, Lactate stimulates fibroblast expression of hyaluronan and CD44: the Warburg effect revisited. *Exp. Cell Res.,* 2002. 276(1): p. 24-31.

[30] Ghatak, S., S. Misra, and B.P. Toole, Hyaluronan oligosaccharides inhibit anchorage-independent growth of tumor cells by suppressing the phosphoinositide 3-kinase/Akt cell survival pathway. *J. Biol. Chem.,* 2002. 277(41): p. 38013-20.

[31] Toole, B.P., T.N. Wight, and M.I. Tammi, Hyaluronan-cell interactions in cancer and vascular disease. *J. Biol. Chem.,* 2002. 277(7): p. 4593-6.

[32] McBride, W.H. and J.B. Bard, Hyaluronidase-sensitive halos around adherent cells. Their role in blocking lymphocyte-mediated cytolysis. *J. Exp. Med.,* 1979. 149(2): p. 507-15.

[33] Zhang, L., C.B. Underhill, and L. Chen, Hyaluronan on the surface of tumor cells is correlated with metastatic behavior. *Cancer Res.,* 1995. 55(2): p. 428-33.

[34] Coussens, L.M. and Z. Werb, Inflammation and cancer. *Nature,* 2002. 420(6917): p. 860-7.

[35] Beckenlehner, K., S. Bannke, T. Spruss, G. Bernhardt, H. Schonenberg, and W. Schiess, Hyaluronidase enhances the activity of adriamycin in breast cancer models in vitro and in vivo. *J. Cancer Res. Clin. Oncol.,* 1992. 118(8): p. 591-6.

[36] Pawlowski, A., H.F. Haberman, and I.A. Menon, The effects of hyalurodinase upon tumor formation in BALB/c mice painted with 7,12-dimethylbenz-(a)anthracene. *Int. J. Cancer,* 1979. 23(1): p. 105-9.

[37] Stern, R., Hyaluronidases in cancer biology. *Semin. Cancer Biol.,* 2008. 18(4): p. 275-80.

[38] Kohno N, Ohnuma T, and T. P., Effects of hyaluronidase on doxorubicin penetration into squamous carcinoma multi- cellular tumor spheroids and its cell letality. *J. Cancer Res. Clin. Oncol.,* 1994. 120: p. 293–297.

[39] Racz GB, Heavner JE, and T. A., Percutaneous lysis of epidural adhesions – evidence for safety and efficacy. *Pain Pract.,* 2008. 8: p. 277-86.

[40] Cavallini, M., R. Gazzola, M. Metalla, and L. Vaienti, The Role of Hyaluronidase in the Treatment of Complications From Hyaluronic Acid Dermal Fillers. *Aesthet Surg. J.,* 2013.

[41] Fraser, S.P., Hyaluronan activates calcium-dependent chloride currents in Xenopus oocytes. *FEBS Lett.,* 1997. 404(1): p. 56-60.

[42] George, J. and R. Stern, Serum hyaluronan and hyaluronidase: very early markers of toxic liver injury. *Clin. Chim. Acta.*, 2004. 348(1-2): p. 189-97.

[43] Evison, M. and V. Wing, Human recombinant hyaluronidase (Cumulase®) improves intracytoplasmic sperm injection survival and fertilization rates. *Reproductive BioMedicine Online*, 2009. 18(6): p. 811-814.

[44] Dunn, A.L., J.E. Heavner, G. Racz, and M. Day, Hyaluronidase: a review of approved formulations, indications and off-label use in chronic pain management. *Expert Opin. Biol. Ther.*, 2010. 10(1): p. 127-31.

[45] Sasson, M. and P. Shvartzman, Hypodermoclysis: an alternative infusion technique. *Am. Fam. Physician,* 2001. 64(9): p. 1575-8.

[46] Gaisford, W. and D.G. Evans, Hyaluronidase in paediatric therapy. *Lancet,* 1949. 2(6577): p. 505-7.

[47] Beaulieu, M.J., Hyaluronidase for extravasation management. *Neonatal Netw.,* 2012. 31(6): p. 413-8.

[48] Spandorfer, P.R., S.E. Mace, P.J. Okada, H.K. Simon, C.H. Allen, D.M. Spiro, et al., A randomized clinical trial of recombinant human hyaluronidase-facilitated subcutaneous versus intravenous rehydration in mild to moderately dehydrated children in the emergency department. *Clin. Ther.,* 2012. 34(11): p. 2232-45.

[49] Rowlett, J., Extravasation of contrast media managed with recombinant human hyaluronidase. *Am. J. Emerg. Med.*, 2012. 30(9): p. 2102 e1-3.

[50] Szabo, L., [Subcutaneous and intramuscular urography with hyaluronidase]. *Orv. Hetil.,* 1954. 95(28): p. 763-4.

[51] Frieling, H., [Urography of infants and small children with subcutaneous injection of the contrast medium supplemented by hyaluronidase]. *Med. Welt.*, 1951. 20(50): p. 1751.

[52] Byrne, J.E. and W.F. Melick, Subcutaneous urography; description of a new method utilizing 70% urokon and hyaluronidase; a preliminary report. *Urol. Cutaneous Rev.*, 1951. 55(4): p. 193-9.

[53] Schulze, C., T. Bittorf, H. Walzel, G. Kundt, R. Bader, and W. Mittelmeier, Experimental evaluation of hyaluronidase activity in combination with specific drugs applied in clinical techniques of interventional pain management and local anaesthesia. *Pain Physician,* 2008. 11(6): p. 877-83.

[54] Dutton, W.A., The use of ergometrine with hyaluronidase to prevent post-partum haemorrhage. *J. Obstet. Gynaecol. Br. Emp.,* 1958. 65(2): p. 315-20.

[55] Kimbell, N., Intramuscular ergometrine and hyaluronidase in prevention of postpartum haemorrhage; use by midwives in 2,002 cases. *Br. Med. J.*, 1954. 2(4880): p. 130-1.

[56] Cohain, J.S., Use of hyaluronidase to prevent perineal trauma during spontaneous delivery: a pilot study. *J. Midwifery Womens Health*, 2009. 54(1): p. 92; author reply 92.

[57] Scarabotto, L.B. and M.L. Riesco, Use of hyaluronidase to prevent perineal trauma during spontaneous delivery: a pilot study. *J. Midwifery Womens Health,* 2008. 53(4): p. 353-61.

[58] Frenzel, K.H., [Protection of the perineum with hyaluronidase]. *Zentralbl. Gynakol.,* 1954. 76(35): p. 1602-4.

[59] Colacioppo, P.M., M.L. Gonzalez Riesco, and M.D. Koiffman, Use of hyaluronidase to prevent perineal trauma during spontaneous births: a randomized, placebo-controlled, double-blind, clinical trial. *J Midwifery Womens Health*, 2011. 56(5): p. 436-45.

[60] Bhavsar, A.R., L.R. Grillone, T.R. McNamara, J.A. Gow, A.M. Hochberg, and R.K. Pearson, Predicting response of vitreous hemorrhage after intravitreous injection of highly purified ovine hyaluronidase (Vitrase) in patients with diabetes. *Invest. Ophthalmol. Vis. Sci.,* 2008. 49(10): p. 4219-25.

[61] Narayanan, R. and B.D. Kuppermann, Hyaluronidase for pharmacologic vitreolysis. *Dev. Ophthalmol.,* 2009. 44: p. 20-5.

[62] Silverstein, S.M., S. Greenbaum, and R. Stern, Hyaluronidase in Ophthalmology. *J. App. Res.,* 2012. 12(1).

[63] Islam, M.N., S. Chakroborty, R. Bandopadhay, and A. Mondal, Sodium bicarbonate versus sodium hyaluronidase in ocular regional anaesthesia--a comparative study. *J. Indian Med. Assoc.,* 2012. 110(1): p. 29-30, 39.

[64] Johnsson, C., R. Hallgren, and G. Tufveson, Hyaluronidase reduces intragraft pressure of rejecting tissue. *Transplant. Proc.,* 2001. 33(4): p. 2484.

[65] Johnsson, C., R. Hallgren, L.O. Lindbom, and G. Tufveson, Edema treatment during cardiac allograft rejection. *J. Heart Lung. Transplant.,* 1999. 18(12): p. 1238-42.

[66] Johnsson, C., R. Hallgren, A. Elvin, B. Gerdin, and G. Tufveson, Hyaluronidase ameliorates rejection-induced edema. *Transpl. Int.,* 1999. 12(4): p. 235-43.

[67] Bertelli, G., D. Dini, G.B. Forno, A. Gozza, S. Silvestro, M. Venturini, et al., Hyaluronidase as an antidote to extravasation of Vinca alkaloids: clinical results. *J. Cancer Res. Clin. Oncol.,* 1994. 120(8): p. 505-6.

[68] Maroko, P.R., P. Libby, C.M. Bloor, B.E. Sobel, and E. Braunwald, Reduction by hyaluronidase of myocardial necrosis following coronary artery occlusion. *Circulation,* 1972. 46(3): p. 430-7.

[69] Kloner, R.A., E. Braunwald, and P.R. Maroko, Long-term preservation of ischemic myocardium in the dog by hyaluronidase. *Circulation,* 1978. 58(2): p. 220-6.

[70] Taira, A., K. Uehara, S. Fukuda, K. Takenada, and M. Koga, Active drainage of cardiac lymph in relation to reduction in size of myocardial infarction: an experimental study. *Angiology,* 1990. 41(12): p. 1029-36.

[71] Sunnergren, K.P. and M.J. Rovetto, Hyaluronidase reversal of increased coronary vascular resistance in ischemic rat hearts. *Am. J. Physiol.,* 1983. 245(2): p. H183-8.

[72] Maclean, D., M.C. Fishbein, P.R. Maroko, and E. Braunwald, Hyaluronidase-induced reductions in myocardial infarct size. *Science,* 1976. 194(4261): p. 199-200.

[73] Szlavy, L., D.F. Adams, N.K. Hollenberg, and H.L. Abrams, Cardiac lymph and lymphatics in normal and infarcted myocardium. *Am. Heart J.,* 1980. 100(3): p. 323-31.

[74] Hillis, L.D., S.F. Khuri, E. Braunwald, R.A. Kloner, D. Tow, E. Barsamian, et al., Assessment of the efficacy of interventions to limit ischemic injury by direct measurement of intramural carbon dioxide tension after coronary artery occlusion in the dog. *J. Clin. Invest.,* 1979. 63(1): p. 99-107.

[75] Wetstein, L., M.B. Simson, J. Haselgrove, C.H. Barlow, and A.H. Harken, Mechanism of action of hyaluronidase in decreasing myocardial ischemia post coronary occlusion in the isolated perfused rabbit heart. *Am. Heart J.,* 1982. 104(3): p. 529-36.

[76] Szlavy, L., S. Kubik, A. de Courten, H.J. Hachen, and F. Solti, CLS 2210, hyaluronidase and the cardiac lymphatic system. *Angiology,* 1985. 36(7): p. 452-7.

[77] Premaratne, S., B.I. Watanabe, W.F. LaPenna, and J.J. McNamara, Effects of hyaluronidase on reducing myocardial infarct size in a baboon model of ischemia-reperfusion injury. *J. Surg. Res.,* 1995. 58(2): p. 205-10.

[78] Cairns, J.A., D.A. Holder, P. Tanser, and E. Missirlis, Intravenous hyaluronidase therapy for myocardial infarction in man: double-blind trial to assess infarct size limitation. *Circulation,* 1982. 65(4): p. 764-71.

[79] Hyaluronidase therapy for acute myocardial infarction: results of a randomized, blinded, multicenter trial. MILIS Study Group. *Am. J. Cardiol.*, 1986. 57(15): p. 1236-43.

[80] Roberts, R., E. Braunwald, J.E. Muller, C. Croft, H.K. Gold, T.D. Hartwell, et al., Effect of hyaluronidase on mortality and morbidity in patients with early peaking of plasma creatine kinase MB and non-transmural ischaemia. Multicentre investigation for the limitation of infarct size (MILIS). *Br. Heart J.*, 1988. 60(4): p. 290-8.

[81] Julian, D.G., J.M. Simpson, P.J. Cadigan, M.C. Petri, R.J. Hall, R.H. Smith, et al., A controlled trial of GL enzyme in the treatment of acute myocardial infarction. *Cardiology,* 1988. 75(3): p. 177-83.

[82] Iwasaki, T., L.G. Ribeiro, D.B. Faria, W.M. Cheung, and P.R. Maroko, Importance of the source of hyaluronidase preparations in determining protective effect on ischemic heart muscle in acute myocardial infarction. *Am. Heart J.*, 1981. 102(3 Pt 1): p. 324-9.

[83] Borrelli, F., F. Antonetti, F. Martelli, and L. Caprino, The co-operative action of hyaluronidase and urokinase on the isoproterenol-induced myocardial infarction in rats. *Thromb. Res.*, 1986. 42(2): p. 153-64.

[84] Iqbal, J., J. Gunn, and P.W. Serruys, Coronary stents: historical development, current status and future directions. *Br. Med. Bull.*, 2013. 106: p. 193-211.

[85] Baring, D. and J. Marshall, How we do it – Hyaluronidase injection for the rhinoplasty patient. *Clin. Otolaryngology*, 2011. 36: p. 588-598.

[86] Narins RS, Coleman WP, and G. RG., Recommendations and Treatment Options for Nodules and Other Filler Complications. *Dermatol. Surg.*, 2009. 35: p. 1667-1671.

[87] Menon H, Thomas M, and D.s. J., Low dose of Hyaluronidase to treat over correction of HA filler - A Case Report. *J. Plas. Reconstr. & Aesth. Surg.*, 2010. 63: p. 416-417.

[88] Christensen, L.H., Host tissue interaction, fate, and risks of degradable and nondegradable gel fillers. *Dermatol. Surg,.* 2009. 35 Suppl 2: p. 1612-9.

[89] Grunebaum LD, Allemann IB, Dayan S, Mandy S, and B. L., The Risk of Alar Necrosis Associated with Dermal Filler Injection. *Dermatol. Surg.*, 2009. 35: p. 1635–1640.

[90] Kim, D.W., E.S. Yoon, Y.H. Ji, S.H. Park, B.I. Lee, and E.S. Dhong, Vascular complications of hyaluronic acid fillers and the role of hyaluronidase in management. *J. Plast. Reconstr. Aesthet. Surg.*, 2011. 64(12): p. 1590-5.

[91] Sclafani, A.P. and S. Fagien, Treatment of injectable soft tissue filler complications. *Dermatol. Surg.,* 2009. 35 Suppl 2: p. 1672-80.

[92] Hirsch RJ, Cohen JL, and C. JD., Successful management of an unusual presentation of impending necrosis following a hyaluronic acid injection embolus and a proposed algorithm for management with hyaluronidase. *Dermatol. Surg.,* 2007. 33(3): p. 357-60.

[93] Menzel, E.J. and C. Farr, Hyaluronidase and its substrate hyaluronan: biochemistry, biological activities and therapeutic uses. *Cancer Lett,* 1998. 131(1): p. 3-11.

[94] Girish, K.S., K. Kemparaju, S. Nagaraju, and B.S. Vishwanath, Hyaluronidase inhibitors: a biological and therapeutic perspective. *Curr. Med. Chem.,* 2009. 16(18): p. 2261-88.

[95] Ahluwalia, H.S., A. Lukaris, and C.M. Lane, Delayed allergic reaction to hyaluronidase: a rare sequel to cataract surgery. *Eye* (Lond), 2003. 17(2): p. 263-6.

[96] Kim, J.H., G.S. Choi, Y.M. Ye, D.H. Nahm, and H.S. Park, Acute urticaria caused by the injection of goat-derived hyaluronidase. *Allergy Asthma Immunol. Res.,* 2009. 1(1): p. 48-50.

[97] Kim, T.W., J.H. Lee, K.B. Yoon, and D.M. Yoon, Allergic reactions to hyaluronidase in pain management -A report of three cases. *Korean J. Anesthesiol.,* 2011. 60(1): p. 57-9.

[98] Szepfalusi, Z., I. Nentwich, M. Dobner, K. Pillwein, and R. Urbanek, IgE-mediated allergic reaction to hyaluronidase in paediatric oncological patients. *Eur. J. Pediatr.,* 1997. 156(3): p. 199-203.

[99] Ebo, D.G., S. Goossens, F. Opsomer, C.H. Bridts, and W.J. Stevens, Flow-assisted diagnosis of anaphylaxis to hyaluronidase. *Allergy,* 2005. 60(10): p. 1333-4.

[100] Leibovitch, I., D. Tamblyn, R. Casson, and D. Selva, Allergic reaction to hyaluronidase: a rare cause of orbital inflammation after cataract surgery. *Graefes Arch. Clin. Exp. Ophthalmol.,* 2006. 244(8): p. 944-9.

[101] Eberhart, A.H., C.R. Weiler, and J.C. Erie, Angioedema related to the use of hyaluronidase in cataract surgery. *Am. J. Ophthalmol.,* 2004. 138(1): p. 142-3.

[102] Borchard, K., R. Puy, and R. Nixon, Hyaluronidase allergy: a rare cause of periorbital inflammation. *Australas J Dermatol,* 2010. 51(1): p. 49-51.

[103] Escolano, F., N. Pares, I. Gonzalez, J. Castillo, A. Valero, and B. Bartolome, Allergic reaction to hyaluronidase in cataract surgery. *Eur. J. Anaesthesiol.,* 2005. 22(9): p. 729-30.

[104] Minning, C.A., Jr., Hyaluronidase allergy simulating expulsive choroidal hemorrhage. *Arch. Ophthalmol.,* 1994. 112(5): p. 585-6.

[105] Lyall, D.A., M. McQueen, K. Ramaesh, and C. Weir, A sting in the tale: cross reaction hypersensitivity to hyaluronidase. *Eye* (Lond), 2012. 26(11): p. 1490.

[106] Stafford, R.S., Regulating off-label drug use--rethinking the role of the FDA. *N. Engl. J. Med.,* 2008. 358(14): p. 1427-9.

[107] Dresser, R. and J. Frader, Off-label prescribing: a call for heightened professional and government oversight. *J. Law Med. Ethics,* 2009. 37(3): p. 476-86, 396.

In: Hyaluronan
Editor: Vitor H. Pomin

ISBN: 978-1-63117-808-5
© 2014 Nova Science Publishers, Inc.

Chapter 5

IN VITRO CHARACTERIZATION AND OPHTHALMIC APPLICATIONS OF HYALURONAN

X. Michael Liu[1], Patricia S. Harmon[1],
E. Peter Maziarz[1] and John W. Sheets[2]*

[1]Global Research & Development, Bausch & Lomb, Rochester, NY, US
[2]Global Research & Development, Anika Therapeutics, Inc.,
Bedford, MA, US

ABSTRACT

Hyaluronan (HA), also known as hyaluronic acid or sodium hyaluronate, has drawn considerable interest and attention from ophthalmic surgical and eye care industries due to its biocompatibility, lubricity, and viscoelastic properties. A more complete understanding of HA biopolymers has therefore become increasingly critical since thorough characterization of raw materials and finished ophthalmic products helps promote product quality and process control. Often such detailed information requires the use of a combination of analytical techniques. In this chapter, we compare size exclusion chromatography (SEC) with on-line multi-angle light scattering (SEC-MALS) and SEC with triple detection (SEC-TD) experiments for HA analysis. Three lots of commercially available eye drop grade HA raw materials were

---

* Email address: Patricia.Harmon@bausch.com

characterized by SEC-MALS and SEC-TD. The absolute molecular weight averages, molecular weight distribution, radius of gyration, intrinsic viscosity, and solution conformation of the HA lots were determined and compared by the two techniques.

In addition, the molecular weights and concentrations of HA in eleven marketed ophthalmic products were evaluated by hyaluronidase digestion experiments and by SEC-TD. The weight-average molecular weights (Mw) of the products tested ranged from 155,000 to 1,400,000 Daltons and the concentration of HA ranged from 0.003% to 0.15%. The work presented in this chapter is from the first publication on characterizing HA molecular weights and concentrations in various marketed HA containing ophthalmic products. Research has indicated that there is a correlation between the molecular weight of hyaluronan and its biocompatibility/biological functions, with high molecular weight HA exhibiting many biological and physiological benefits. Future investigation of the effect of low molecular weight HA on eye is required.

1. INTRODUCTION

Hyaluronan (HA), also known as hyaluronic acid or sodium hyaluronate, is a naturally occurring, non-branched polysaccharide that consists of alternately repeating monosaccharide units of D-glucuronic acid and N-acetyl D-glucosamine linked by β 1-3 and β 1-4 glycosidic bonds. The molecular structure is shown in Figure 1.

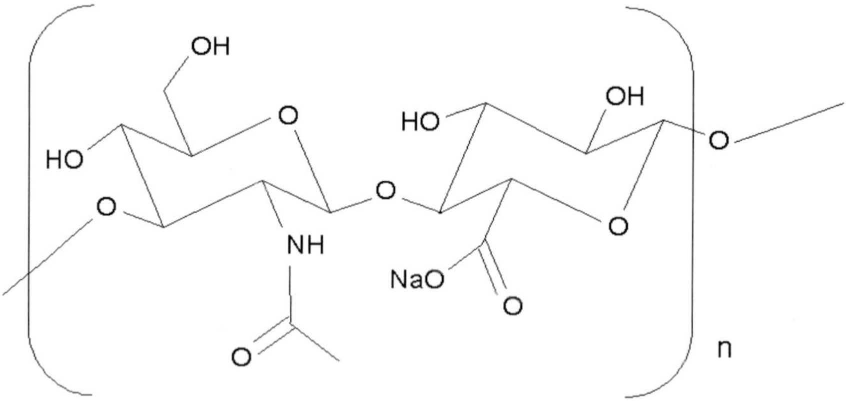

Figure 1. Molecular structure of hyaluronan.

HA is existent to some extent in virtually all biological fluids and tissues [1]. Specifically, it is a common component of synovial fluid, the extracellular matrix of connective tissue in the skin, the vitreous humor, and the tear fluid. This biopolymer is a natural lubricant and is extremely hygroscopic, absorbing up to 1,000 times its own weight in water when hydrated. HA enhances tear stability and retards tear removal due to its viscoelasticity [2], and promotes corneal wound healing by stimulating corneal epithelial cell migration [3-6]. It has also been found to play a significant role in reducing protein deposition on contact lens materials [7]. Furthermore, HA is thought to have a protective effect against oxidative damage to cells by inhibiting free radicals [8-10]. The protective effect appears to increase with increasing HA concentrations [8]. It is hypothesized that HA also increases corneal wettability due to the enhanced water retention on the corneal surface, making it useful for the treatment of dry eye [11].

Its unique physical and chemical characteristics and properties make HA an appealing natural biopolymer to the healthcare industry in general, including the ophthalmic industry [7, 12, 13]. The typical marketed ophthalmic products include artificial tears, contact lenses, multipurpose lens care solutions, and rewetting eye drops [14]. HA has been added to contact lens products to enhance the wettability and comfort of contact lenses [15]. HA is also used in ophthalmologic surgeries. In cataract surgery, it is used as a viscoelastic to raise the intraocular pressure; and as a coating on PMMA-based intraocular lenses in order to reduce protein and bacterial adhesion onto the lens surfaces [12].

The family of HA biopolymers covers a very broad molecular weight range up to several millions Daltons. [16] Research has found that the function and effect of HA varies significantly based on its molecular weight. [17] High molecular weight HA hydrates tissues, promotes tissue integrity, lubricates, and is space-filling and shock-absorbing. It also impedes differentiation, and is anti-angiogenic, anti-inflammatory, and immunosuppressive. In contrast, lower molecular weight HA tends to send "danger signals", and is inflammatory, immuno-stimulatory, and angiogenic. [18-20]

The HA found in human tear film typically has a molecular weight over 1 million Daltons. [21, 22] HA is known to degrade into smaller molecular weight fragments in aqueous solutions over time. Research has indicated that there is a correlation between the molecular weight of HA and its biocompatibility/biological functions in ocular tissues, osteoarthritis, and general biological research [14, 21, 23-27]. Given this correlation between the molecular weight of HA and its biological functions, molecular weight should be an important considering factor when developing ophthalmic products. It is essential to use high molecular weight HA as a starting material in ophthalmic products. This will help ensure that a certain molecular weight of HA is maintained over the shelf life of the product, which will help enhance comfort.

Size exclusion chromatography (SEC) and field flow fractionation (FFF) with various advanced detection methods prove to be effective tools to evaluate the solution properties of HA and other biomaterials to ensure the quality of the products for specific claims. [28, 29] We have compared on-line size exclusion chromatography with multi-angle light scattering (SEC-MALS) [27] and SEC with triple detection (SEC-TD) [31, 32] experiments for HA raw material analysis. We have also characterized and compared the molecular weights, molecular weight distributions, and concentrations of HA present in a series of commercially available HA-containing ophthalmic products.

2. EXPERIMENTAL

2.1. Materials

Two sodium hyaluronate (HA) samples (HA-1 and HA-2) were obtained from Bloomage Freda Biopharm Co. (Jinan, China), while HA-3 was obtained from Novozymes (Bagsvaerd, Denmark). A narrow molecular weight poly(ethylene oxide) standard with M_p of 258,000 g/mol ($M_w/M_n = 1.12$) was purchased from American Polymer Standards Corporation (Mentor, OH, USA). Bovine serum albumin was purchased from Sigma (St. Louis, MO, USA). Sodium chloride and glacial acetic acid were obtained from J.T. Baker (Phillipsburg, NJ, USA). Triethylamine hydrochloride was obtained from TCI America (Portland, OR, USA), and sodium azide from EMD Chemicals (Darmstadt, Germany). Borate buffered saline was made in-house at Bausch & Lomb R&D laboratories (Rochester, NY, USA).

Biotrue™ MPS (Bausch & Lomb, Rochester, NY, U.S.A.), Sodyal MPS (Omisan Pharmaceuticals, Italy), Avizor Unica Sensitive MPS (Avizor, Spain), Hy-Care MPS (Sauflon Pharmaceuticals Ltd.,UK), Hyalu-Sol MPS (Italy), Blink Contact Lubricating Eye Drops (Abbott Medical Optics, Santa Ana, CA, U.S.A), Aquify Long-Lasting Comfort Drops (Ciba Vision, Duluth, GA, U.S.A.), Safigel contact lens packaging solution (Safilens, S.R.L., Staranzano, Italy), Weicon Brand B MPS (Weicon Optics Ltd. Shanghai, China), Eye Secret MPS (Hydron Optics Ltd. Taiwan), and Simply One MPS

(Perret Opticiens, Carouge/Geneva, Switzerland) are commercially available HA-containing ophthalmic products.

2.2. Chromatographic Conditions

HA raw material samples were prepared at 0.25 mg/mL in borate buffered saline (BBS), and were allowed to swell for 72 hours for the HA to become completely solvated prior to analysis. The separation of these samples was achieved using one Shodex SB-807HQ column and one Shodex SB-806HQ column, in series. The columns were kept in a column oven at ambient temperature. The mobile phase used was 0.15M NaCl + 0.02% NaN_3, and was filtered using a 0.02 μm Anodisc™ filter prior to use in order to help minimize the noise level in the light scattering traces. The flow rate was set at 0.5 mL/min, and 100 μL of the sample solutions were injected for each analysis. The dn/dc value of an aqueous solution of HA used for calculation was 0.167 mL/g. [33]

The commercially available HA-containing ophthalmic products were analyzed as received, with the exception of the Blink Contact Lubricating Eye Drops and the Aquify Long-Lasting Comfort Drops, which were diluted 5x and 3x, respectively, with BBS. The separation of HA from the rest of the components in the commercial products was achieved using two Ultrahydrogel Linear (Waters Corporation, Milford, MA, U.S.A.) columns connected in series. These columns are packed with a semi-rigid polymeric gel that is hydroxylated PMMA in nature. They can be expected to have a small amount of free carboxyl groups as a result of hydrolysis, which can interact adversely with components present in the ophthalmic products (i.e. cationic disinfectants and preservatives). The columns were kept in a column oven at 35°C. The mobile phase used was 0.1M triethylamine hydrochloride with 1% acetic acid, and was filtered using a 0.02 μm Anodisc™ filter prior to use in order to help minimize the noise level in the light scattering traces. This mobile phase minimizes interactions since the partially ionized triethylamine effectively caps the anionic sites on the column packing material [31]. The flow rate was set at 0.7 mL/min, and 100 μL of the sample solutions were injected for each analysis. The dn/dc value of an aqueous solution of HA used for calculation was 0.167 mL/g. [33]

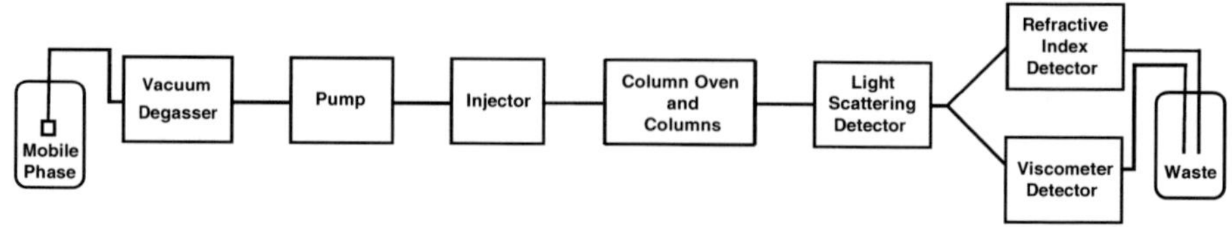

Figure 2. Schematic of size exclusion chromatography with triple detection (SEC-TD).

2.3. SEC with Triple Detection

An Integrated PL-GPC50 Plus instrument (Varian, Palo Alto, CA, USA) was used for the SEC with triple detection (SEC-TD) analysis. This instrument contains a pump for solvent delivery, an oven to house the columns, and three detectors: dual-angle (45° and 90°) light scattering, differential refractometer, and viscometer. Upon eluting off the column, the sample entered the light scattering detector, after which the flow was split 1:1 between the refractometer and viscometer. The schematic of the SEC-Triple Detection system is shown in Figure 2.

The SEC-TD system was calibrated using a 0.5 mg/mL solution of a polyethylene oxide narrow standard (M_p = 258,000 g/mol; M_w/M_n = 1.12) prepared in BBS. This was used for determining the inter-detector delay constants (IDD), as well as for the detector constants. The dn/dc value of the aqueous solution of PEO used for calibration was 0.135 mL/g, as given on the certificate of analysis provided with the standard. Data acquisition and calculations were performed using Cirrus Multi GPC/SEC software version 3.1.

Two-sample t-tests were used to compare the molecular weights of HA in each of the products with the molecular weight of HA in Biotrue™ MPS.

2.3.1. SEC-TD Theory

The following simplified equations reflect the correlations between the detector signals and polymer molecular weight, intrinsic viscosity, and molecular radii for conformation determination:

$$R_{dRI} = K_{RI} \times C \times \frac{dn}{dc} \tag{1}$$

$$R_{visc} = K_{Visc} \times C \times IV \tag{2}$$

$$R_{LS}(\theta) = K_{LS}\, \frac{4\pi^2 n_0^2\, (\dfrac{dn}{dc})^2}{\lambda_0^4 N_A} \times C \times P(\theta) \times M_w \qquad (3)$$

$$IV = k_M \times M_v^a \qquad (4)$$

$$R_g' = 0.328 \times \left([\eta] \times M\right)^{1/3} \qquad (5)$$

$$R_h = 0.251 \times \left([\eta] \times M\right)^{1/3} \qquad (6)$$

where R_{dRI}, R_{visc} and $R_{LS}(\theta)$ are signals of the refractive index detector, viscosity detector and light scattering detector, respectively; K_{RI}, K_{visc} and K_{LS} are the instrument constants of refractive index, viscosity and light scattering detectors respectively. The concentration of the sample solution is denoted by C, and the specific refractive index increment is given by dn/dc. The viscosity and weight-average molecular weights are M_v and M_w, respectively. For an individual fraction from SEC, M_v is equal to M_w since they are treated as monodisperse samples. The intrinsic viscosity of the polymer is represented by IV. The scattering angle is θ and $P(\theta)$ is the form factor, which relates the angular variation in scattering intensity to the mean square radius of the particle. The excess light scattering intensity at the zero angle is calculated from the 90° excess scattering intensity by using the particle size information for $P(\theta)$ calculation under appropriate assumptions [34]. The refractive index of the solvent is denoted as n_0, λ_0 is the wavelength of the light, and N_A is the Avogadro number. Equation 4 is the Mark-Houwink equation, with K_M and exponent a being Mark-Houwink's constants. The value of exponent a (or the conformational coefficient) can also be used to predict the combined effect of polymeric chain conformations and the presence of branching. It is well known that values of $a \geq 1.0$ indicate a rigid rod conformation, values between 0.5 and 0.8 indicate a random coil in a theta and good solvent respectively, and $a \leq 0.5$ indicates a spherical or branching conformation. [35]. Equations 5 and 6 are the formulae used for calculating the radius of gyration (R_g') and hydrodynamic radius (R_h) which are derived from the Einstein viscosity law for particles in suspension under certain assumptions. [31, 34, 36]

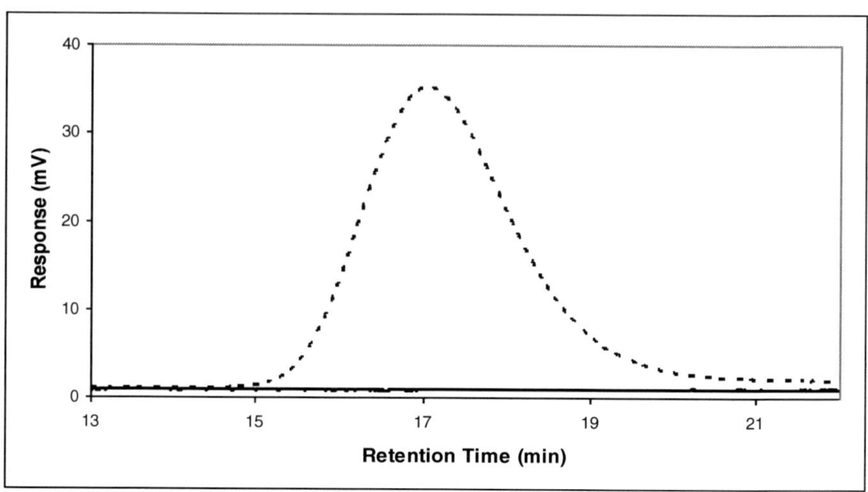

Figure 3. Overlay of viscometer chromatograms of Biotrue™ with and without exposure to Hyaluronidase.

2.3.2. SEC-TD with Enzymatic Process

SEC-TD reveals the presence of polymer materials without providing the positive identification (i.e. chemical structure, end group chemistry, and molecular composition/formula). Therefore, we employed an orthogonal test to increase the confidence and confirm the presence of HA in the polymeric peak analyzed in each of the eleven ophthalmic products tested. Hyaluronidase is an enzyme known to digest or break down of HA molecules *specifically*. Each of the eleven products was analyzed by SEC-TD both before and after exposure to an aqueous solution of hyaluronidase. In order to have complete reaction, we allowed the digestion experiments to go over night. If a peak in the chromatogram were entirely constituted of HA, that peak would have completely disappear (shifted to the solvent front or the very low molecular weight region where we do not monitor in the experiment) after the enzymatic treatment. If the polymeric peak in the chromatogram of a given product had the same intensity before and after the hyaluronidase treatment, it was concluded that HA was not detectable in that product using the chromatographic method described in this study. Finally, if the polymeric peak was present but with a decrease in intensity after the enzymatic treatment, it was concluded that there was at least other polymeric component in the ophthalmic product co-eluting with HA. This interference hindered the ability to accurately determine the concentration and molecular weights of HA in this product in the same way used for the other products. The three scenarios described above were observed in Biotrue™, Eye Secret, and Simply One, respectively (Figures 3-5).

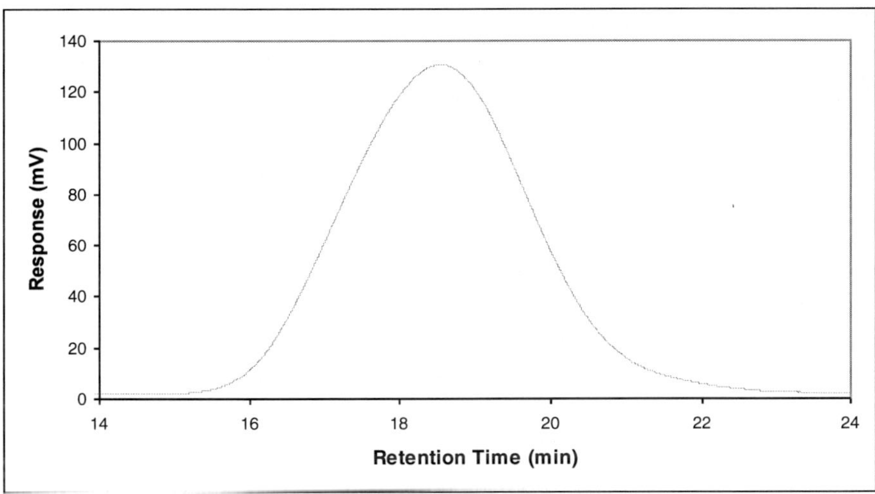

Figure 4. Viscometer chromatograms of Eye Secret with and without exposure to Hyaluronidase.

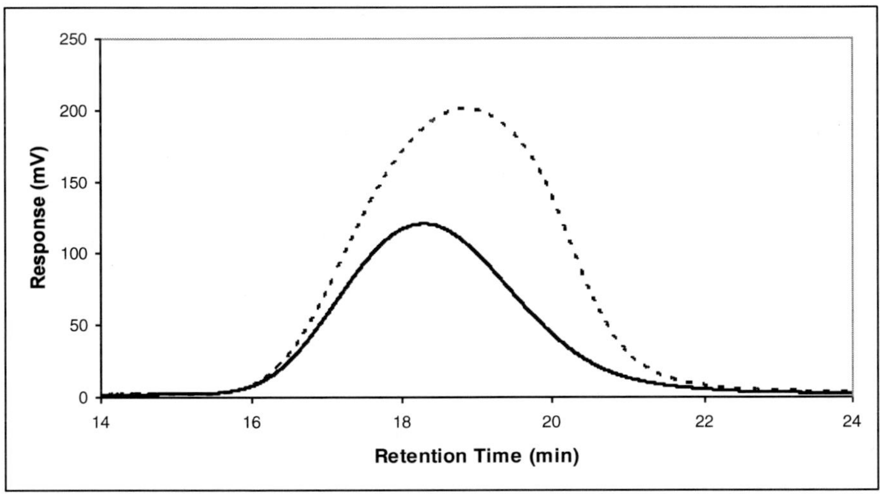

Figure 5. Overlay of viscometer chromatograms of Simply One with and without exposure to Hyaluronidase.

2.4. SEC-MALS

A pump and an autosampler from the Agilent 1100 series (Agilent Technologies, Santa Clara, CA, USA) were utilized for the SEC-MALS analysis. An Optilab rEX and HELEOS II (Wyatt Technologies, Santa Barbara, CA USA) were used as the concentration and MALS detectors, respectively. The HELEOS II utilizes a 120mW solid-state laser at 658 nm, and in aqueous systems, measures the intensity of scattered light at 17 angles. The calibration constant was calculated using toluene. The Wyatt Technology software ASTRA assumes a Rayleigh factor of 1.1705×10^{-5} cm^{-1} for toluene at 658 nm. A Bovine serum albumin (BSA) solution in mobile phase was used to determine the inter-detector delay volumes, band broadening parameters, and normalization coefficients. Data acquisition and calculations were performed using ASTRA software version 5.3. The Zimm model with a 1st order fit and an A_2 value of 4×10^{-3} mol mL/g^2 were used for the HA samples. The A_2 value used was based on single-batch experiments.

2.4.1. SEC-MALS Theory

Static light scattering is a well-known technique for dilute solution characterization. Light scattering can be used to determine the absolute molecular weight and molecular size according to equation 7:

$$\frac{Kc}{R_\theta} = \frac{1}{P(\theta)M} + 2A_2c \tag{7}$$

where the constant K and form factor $P(\theta)$ are given as:

$$K = \frac{4\pi^2 n_0^2 (dn/dc)^2}{\lambda_0^4 N_A}$$

$$P(\theta)^{-1} = 1 + \frac{16\pi^2 \langle R_g^2 \rangle}{3\lambda_0^2} \sin^2(\theta/2)$$

and c_i is the concentration of the polymer, R_θ is the excess Rayleigh factor, θ is the scattering angle, n_0 is the refractive index of the solvent, λ_0 is the wavelength of the light, and N_A is the Avogadro number. [37, 38]

When light scattering is combined with SEC, Debye plots are obtained for each SEC fraction, or slice, that elutes off the column. The refractive index detector determines the concentration of each fraction, c_i. Each Debye plot can be used to determine the corresponding M_i values. By combining the concentration and molecular weight for each slice of the peak, molar mass moments are determined, as defined below:

$$M_w = \frac{\sum n_i M_i^{\,2}}{\sum n_i M_i} \tag{8}$$

$$M_n = \frac{\sum n_i M_i}{\sum n_i} \tag{9}$$

where M_w and M_n represent weight-average molecular weight and number-average molecular weight of the polymer sample, respectively. The number of polymer molecules at molecular weight M_i is given by n_i. The molecular weight distribution, or polydispersity (*PD*), can be determined according to equation 10.

$$PD = \frac{M_w}{M_n} \tag{10}$$

3. RESULTS AND DISCUSSION

3.1. Molecular Weight of HA Raw Materials

The average molecular weight values (M_w, M_n) and molecular weight distribution (PD) of three lots of HA raw materials obtained by SEC-TD and SEC-MALS are given in Table 1. Each reported value is the mean result of six measurements. There is good agreement between the two analytical techniques for both the weight-average and number-average molecular weight values. The average M_w values are within 5% and the average M_n values are within 12% of each other for each of the three lots of HA tested. The M_w and M_n values calculated by both techniques are also consistent with literature values for the HA raw materials investigated [39, 40].

Table 1. Comparison of weight-average molecular weight (M_w), number-average molecular weight (M_n), and molecular weight distribution (PD) results for HA raw materials obtained from SEC-TD and SEC-MALS

HA Lot	M_w (10^5 Da)	M_n (10^5 Da)	PD	Technique
HA-1	15.85 ± 0.63	9.88 ± 0.94	1.61 ± 0.11	SEC-TD
	15.19 ± 0.44	11.13 ± 0.32	1.36 ± 0.01	SEC-MALS
HA-2	8.38 ± 0.18	5.63 ± 0.16	1.49 ± 0.02	SEC-TD
	8.55 ± 0.23	6.21 ± 0.15	1.38 ± 0.02	SEC-MALS
HA-3	6.42 ± 0.17	4.55 ± 0.17	1.41 ± 0.03	SEC-TD
	6.36 ± 0.16	4.57 ± 0.20	1.39 ± 0.03	SEC-MALS

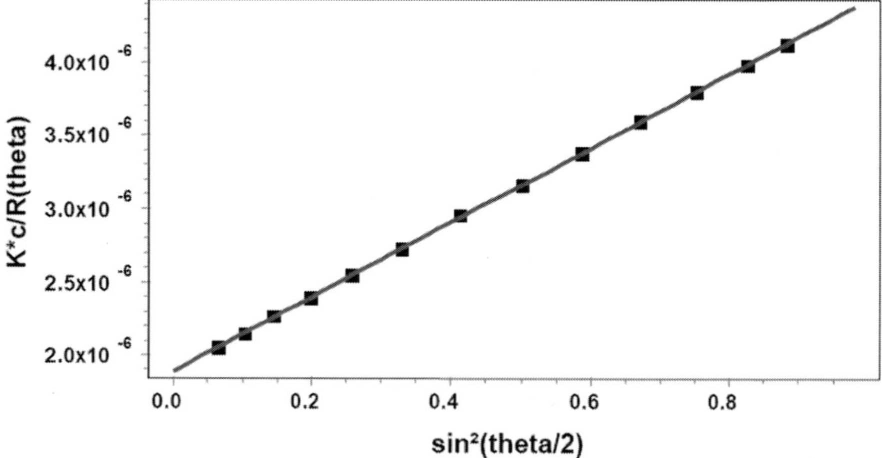

Figure 6. Debye plot, $K*c/R(\theta)$ vs. $\sin^2(\theta/2)$, for a SEC fraction of HA-2 using SEC-MALS.

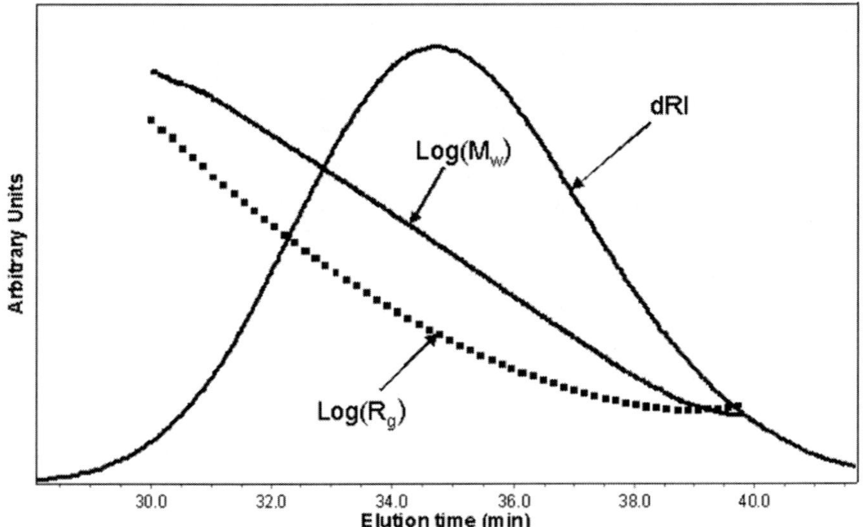

Figure 7. Distributions of molecular weight and molecular size of HA-2 using SEC-MALS.

Figure 6 contains the Debye plot of one SEC fraction of HA-2. The Debye plot shows the extrapolation to the zero angle for one of the HA fractions in the SEC-MALS chromatogram. Using the concentration of the HA fraction based on the intensity of the on-line refractive index detector, the molecular weight and the molecular size of this HA fraction can be calculated. Combining the data from each of the fractions across the HA peak in the SEC chromatogram, we are able to determine the weight-average and number-average molecular weight of the polymer (Equations 8-9). The resulting molecular weight distribution for HA-2 is shown in Figure 7.

3.2. Solution Properties of HA

Solution properties (radius of gyration, intrinsic viscosity, and conformation) of three lots of HA raw materials were also determined. The radii of gyration determined for the three lots of HA raw materials are summarized in Table 2. Each reported value is the mean result of six measurements. The average molecular radii of gyration (R_g) obtained using SEC-MALS were 23-35% higher than the R_g' values obtained using SEC-TD. R_g' is calculated by SEC-TD from intrinsic viscosity and 2-angle light scattering (i.e. 45 and 90 degrees) as shown in Equation 5. The M is calculated by a single-point two-angle light scattering measurement using SEC-TD. Therefore the R_g' value by SEC-TD is less reliable because it does not fully correct the angular dependence effect of large molecules in the solution phase. By contrast, using a multi-angle light scattering detector should reduce the angular dependence effect and therefore provide more reliable data than the dual-angle light scattering detector used in SEC-TD [41-43]. Also, according to the instrument instruction manual and technical support from Polymer Laboratories /Agilent, the R_g' values determined by SEC-TD are based on many assumptions and extrapolations [36]. By contrast, the R_g values determined by SEC-MALS are directly measured and calculated from multiple light scattering angles and thus are believed to be more accurate [37, 38]. The molecular radii of gyration (R_g) determined in the SEC-MALS experiments are consistent with historic literature values for HA with similar molecular weights and matrices [39, 44, 45].

Intrinsic viscosity (IV) is a measure of the occupied hydrodynamic volume of a polymer molecule in a dilute solution and is therefore a reflection of the polymer size. The value of IV can be obtained from the Mark-Houwink equation exhibited in the SEC-TD theory section (Equation 4). In our laboratory configuration, the SEC-MALS system was not able to measure intrinsic viscosity. The IV results determined in this study from SEC-TD (Table 2) are in reasonable agreement with the values in the previously published literature [45, 46].

Table 2. Hyaluronan solution properties obtained from SEC-TD and SEC-MALS

HA Lot	R_g' (nm)	R_g (nm)	IV (dL/g)	Slope of $\log R_g$ vs. $\log M_w$	Technique
HA-1	117.3 ± 1.6	N/A	22.98 ± 1.60	N/A	SEC-TD
	N/A	179.0 ± 2.8	N/A	0.60 ± 0.01	SEC-MALS
HA-2	90.9 ± 1.5	N/A	16.46 ± 0.57	N/A	SEC-TD
	N/A	128.9 ± 2.5	N/A	0.60 ± 0.01	SEC-MALS
HA-3	81.7 ± 4.4	N/A	14.27 ± 0.34	N/A	SEC-TD
	N/A	106.1 ± 1.1	N/A	0.58 ± 0.01	SEC-MALS

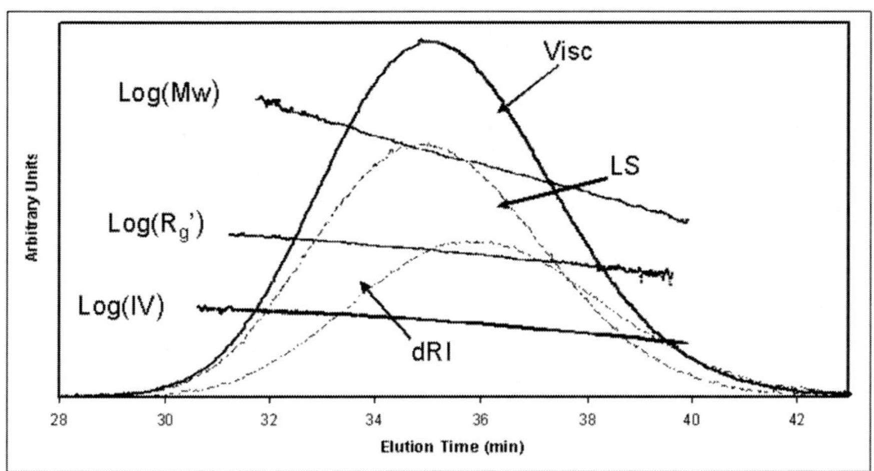

Figure 8. The light scattering (LS), differential refractive index (dRI), and viscometry (Visc) chromatograms of HA-2 superimposed with distributions of molecular weight, radius of gyration, and intrinsic viscosity by SEC-TD.

The molecular weight, radius of gyration, and intrinsic viscosity distributions along with three chromatographic traces for HA-2 are shown in Figure 8. The light scattering (LS), differential refractive index (dRI), and viscometry (Visc) traces illustrate well-behaved peak shapes under test conditions. When comparing the peak intensities and signal-to-noise ratio (S/N) from the three detectors, it was found that the viscometer gave the most intense signal and lowest S/N, while the differential refractive detector gave the lowest signal intensity and highest S/N. As expected, the molecular weight (M_w), radius of gyration (R_g'), and intrinsic viscosity (IV) of HA decreases with increasing elution time. The intrinsic viscosity and intrinsic viscosity distribution are important attributes for better understanding rheological characteristics of HA in test solutions.

Due to the non-linear nature of the Mark-Houwink plots of all three HA samples shown in Figure 9, it is not practical to fit the data to a straight line. The second order polynomial fit function was used for analyzing the data [47]. The correlation coefficients (R^2) for all three samples were greater than 0.99 suggesting the 2nd order polynomial fits are satisfactory. The second order polynomial fits for the three respective HA samples are expressed in equations 11-13.

HA-1: $\log(IV) = -0.4532(\log M_w)^2 + 5.9034(\log M_w) - 17.7656$
$$R^2 = 0.9954 \tag{11}$$

HA-2: $\log(IV) = -0.3241(\log M_w)^2 + 4.3920(\log M_w) - 13.3785$
$$R^2 = 0.9961 \tag{12}$$

HA-3: $\log(IV) = -0.2116(\log M_w)^2 + 3.1333(\log M_w) - 9.8708$
$$R^2 = 0.9977 \tag{13}$$

The Mark-Houwink exponent a value at each SEC elution fraction slice was calculated by taking the derivative of the respective function. The average a values for each of the three HA lots were determined to be between 0.5 and 0.6 at 1.0M g/mol and approximately 0.7 at 500K g/mol. These values are in reasonably good agreement with previously published data [16, 39, 45, 48].

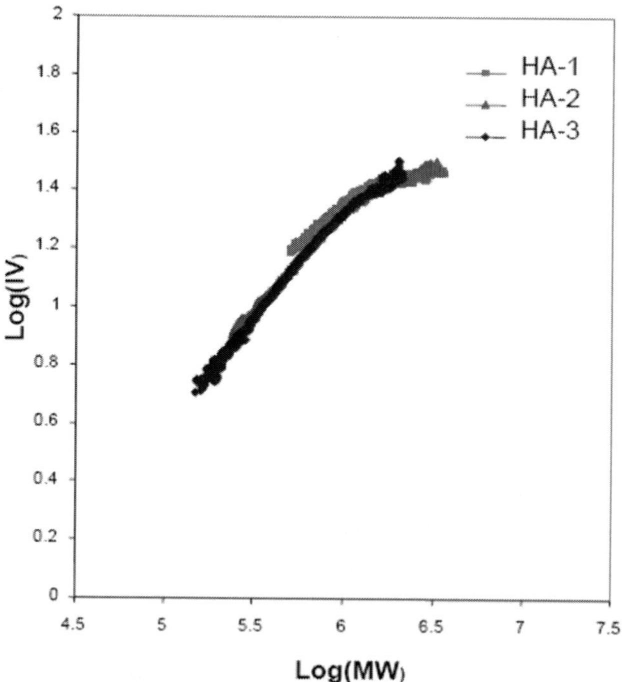

Figure 9. Mark-Houwink plot of three lots of HA.

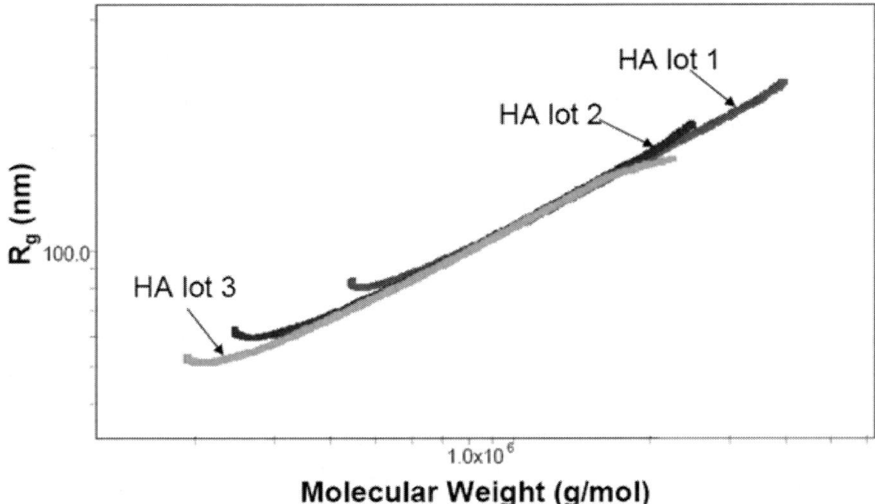

Figure 10. Conformation plots for three lots of HA from SEC-MALS experiments.

Similar to the Mark-Houwink exponent a values determined in SEC-TD experiments, SEC-MALS experiments provide the slope of R_g versus M_w (i.e. conformation plots). Figure 10 shows the conformation plots, log R_g versus log M_w, for all three HA samples from SEC-MALS experiments. The slope of this plot can also be used to determine polymer conformation in solutions [49]. For the three lots of HA tested, the values of the conformation plot slope are 0.6, which is consistent with HA having a random coil conformation in the aqueous solution tested. This result is in good agreement with the general findings in literature reports [28, 29, 40, 50].

3.2. Analysis of HA in Ophthalmic Products

The average molecular weight values (M_w, M_n) and molecular weight distributions (PD) of HA in each of the eleven ophthalmic products tested were obtained by SEC-TD according to Equation 3. The values are given in Table 3. Each reported value is the mean result of a minimum of three measurements.

Table 3. Molecular Weight, Molecular Weight Distribution (PD), and Calculated Concentration of HA in Commercially Available Ophthalmic Products

Sample	Mw (Daltons)	p (T<=t) two-tail	Mn (Daltons)	p (T<=t) two-tail	Calc. HA Conc.	PD M_w/M_n	# Lots; # Total Measurements
Biotrue™ MPS (U.S.A.)	1,400,000	N/A	927,000	N/A	0.01%	1.51	22 lots; >50 measurements
Sodyal MPS (Italy)	461,000	1.88E-12	295,000	1.85E-10	0.003%	1.67	1 lot; 8 measurements
Avizor Unica Sensitive MPS (Spain)	705,000	6.53E-20	511,000	4.63E-14	0.01%	1.39	1 lot; 10 measurements
Hy-Care MPS (UK)	681,000	3.46E-21	463,000	2.20E-15	0.01%	1.47	4 lots; 18 measurements
Hyalu-Sol (Italy)	390,000	2.93E-26	293,000	9.60E-21	0.01%	1.34	1 lot; 7 measurements
Blink Contact Lubricating Eye Drops, (U.S.A.)	624,000	6.42E-22	392,000	3.20E-19	0.15%	1.60	4 lots; 11 measurements
Aquify Long-Lasting Comfort Drops, (U.S.A.)	739,000	2.19E-10	500,000	7.73E-09	0.10%	1.49	3 lots; 8 measurements
Safigel packaging solution (Italy)	155,000	2.34E-28	99,000	1.77E-25	0.08%	1.57	1 lot; 6 measurements
Weicon Brand B MPS (China)	HA was not detected using this SEC-TD method						
Eye Secret MPS (Taiwan)	HA was not detected using this SEC-TD method						
Simply One MPS (Switzerland)	HA was detected but could not be quantitated using this SEC-TD method						

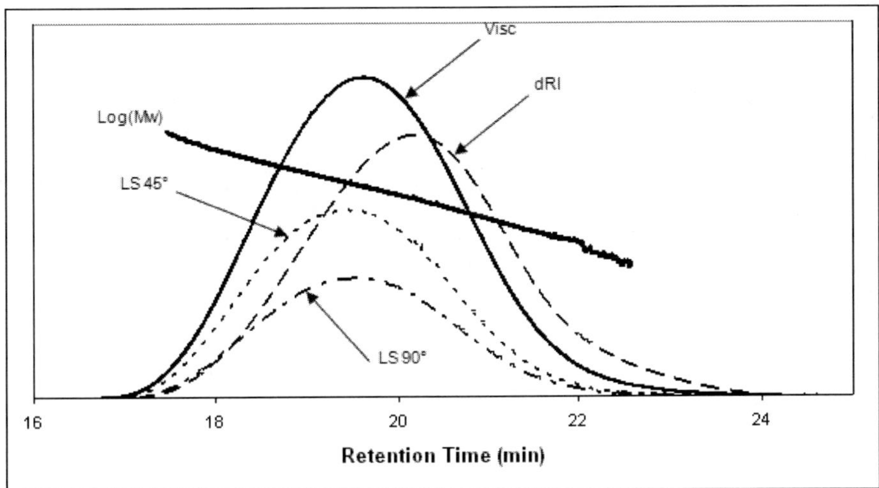

Figure 11. Molecular weight distribution of HA using aqueous SEC-TD in Safigel contact lens packaging solution.

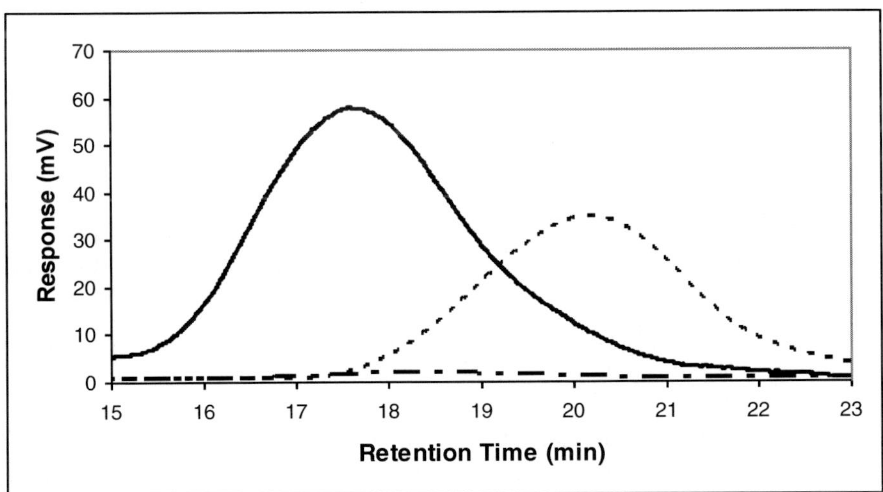

Figure 12. Overlay of dRI chromatograms of three HA-based ophthalmic products (Blink Contact Lubricating Eye Drops, Sodyal MPS, Safigel Contact Lens Packaging Solution) tested in the study. The molecular weight distribution of HA for one of the ophthalmic products tested (Safigel contact lens packaging solution) is shown in Figure 11. The overlay of differential refractive index (dRI) chromatograms of three ophthalmic products (Sodyal MPS, Weicon Brand B

MPS, and Aquify Long-Lasting Comfort Drops) is shown in Figure 12. These are three ophthalmic products with HA molecular weights ranging from 280,000 Daltons (Da) to 739,000 Da, and HA concentrations ranging from 0.003% to 0.15%. An earlier retention time for the refractive index peak indicates a higher molecular weight. Refractive index peak area is directly correlated with the concentration of HA present in the sample.

The weight-average molecular weights (M_w) of the products tested ranged from 155,000 Da to 1,400,000 Da and the number-average molecular weight (M_n) of the products ranged from 99,000 Da to 927,000 Da. When the molecular weights of the solutions were evaluated and compared, the molecular weights of HA in the other ophthalmic products reported were determined to be statistically lower than those of HA in Biotrue™ MPS ($p<0.05$ for all comparisons). The polydispersity (PD $=M_w/M_n$) values of most of the samples tested were between 1.4 and 2.0, which is consistent with the PD of raw materials [51].

The concentration of HA was determined by comparing the differential refractive index (dRI) detector response of the HA peak in each solution with the response of a solution containing 0.01% (w/w) HA. The dRI detector is considered to be a universal detection system in which HA with various molecular weights exhibits similar refractive index values. Therefore the dRI response is directly proportional to the concentration of HA regardless of its molecular weight.

The concentration of HA in the eleven ophthalmic products tested ranged from 0.003% to 0.15%, as shown in Table 3. In general, the concentrations in both the contact lens packaging solution and eye drops tested were higher than those in the multi-purpose solutions. Both the concentrations and the average molecular weights of HA vary significantly in different MPS solutions. Each MPS solution contains cationic disinfectants /preservatives such as polyhexamethylene biguanide (PHMB) and/or quaternary amines (i.e. polyquaternium-1, PQ-1) for disinfecting contact lenses. It is well-known that the anionic HA will interact with these cationic disinfectants. Therefore, it is reasonable to believe that there are certain reaction equilibria between HA and other cationic species for the formulation/product development process and product stability perspectives.

A wide range of HA molecular weights and concentrations were found in the eleven ophthalmic products characterized in this study, with the HA found in Biotrue™ MPS having the highest molecular weight ($p<0.05$). Many ophthalmic product manufacturers use heat or autoclave sterilization in the manufacturing of their products. However, HA is degraded by high temperatures, and therefore the use of heat sterilization may result in ophthalmic products with lower molecular weight HA. The recently granted patent (Sterile HA solutions) describes the challenging and novel manufacturing process used to preserve the high molecular weight of HA during the sterilization process for Biotrue™ MPS [52].

The categorization of the molecular weight of HA is important as higher molecular weight HA has been shown to exhibit many biological benefits [21, 23-27] while lower molecular weight HA has shown complex biological responses which are potentially less favorable to the eye [23]. High molecular weight HA is believed to be larger than 800,000 Daltons [14, 21, 23-25]. Anything below 800,000 Daltons are considered to be medium and low molecular weight HA.

The biological and physical benefits of high molecular weight HA are reported to include protection of corneal [24, 25] and conjunctival epithelium [24], and anti-inflammation [24, 27]. Pauloin et al (2008) investigated the in vitro ability of three different molecular weights of HA (1,000,000 Daltons, 100,000 Daltons, and 20,000 Daltons), to reduce the preservative toxicity of benzalkonium chloride (BAK) on human epithelial models [25]. Testing of two human epithelial cell lines showed that HA 1,000,000 Daltons significantly reduced all of the BAK-induced cytotoxic effects observed in the study. Specifically, HA 1,000,000 Daltons decreased oxidative stress, BAK-induced mitochondrial mass, chromatin condensation and plasma-membrane permeability, and DNA fragmentation of conjunctival cells. In addition, HA 1,000,000 Daltons protected the actin structure against BAK-induced cytoskeleton disorganization. The cytoprotective effects of HA observed in the study decreased with the decreasing molecular weight of HA. The authors concluded that only the high molecular weight HA had a significant cytoprotective effect.

Pauloin et al (2009) have also shown that high molecular weight HA had significant protective effects against UVB radiation [24]. More specifically, their results indicated that high molecular weight HA significantly decreased UVB-induced cell death, significantly decreased UVB-induced caspases-3 and -8 activation but not caspase-9 activation, and significantly decreased UVB-induced IL-6 and IL-8 production. No protective effect was found for UVB-induced oxidative stress or UVB-induced DNA damage and p53 activation. The authors concluded that high molecular weight HA provided anti-inflammatory and anti-apoptotic signals to cells exposed to UVB radiation.

Guillaumie et al (2010) studied the influence of HA molecular weight on its water binding capacity, rheological properties, precorneal residence time in rabbits, and the tolerance of ophthalmic solutions [14]. The authors tested two types of hyaluronic acid: medium molecular weight HA from *Bacillus*-fermentation (Mw = 680,000 Daltons, 770,000 Daltons, 890,000 Daltons, or 1,140,000 Daltons) and high molecular weight HA from *Streptococcus*-fermentation (Mw = 1,5000,000 Daltons or 2,250,000 Daltons). Their results showed that all HA samples bound very high amounts of water and the amount of bound water was independent of molecular weight and the origin of the HA. The kinematic viscosity was strongly dependent on the molecular weight and the concentration of HA. The kinematic viscosity of commercial eye drops typically varies from 2 to 7 centipoise [14] which can be met using both medium and high molecular weight HA depending on the concentration of HA used. A prolonged residence time in the precorneal area was observed for a 0.3% solution of high molecular weight HA compared to a medium molecular weight solution at the same concentration. In vivo ocular tolerance was noted for both types of HA after topical installation onto the corneal surface. The authors noted that medium molecular weight HA is superior to high molecular weight HA due to the ease of sterilization by filtration and manufacturing of medium weight HA. However, techniques for sterilization of high molecular weight HA have been developed to overcome this concern [52].

Kikuchi et al (1996) showed that when comparing medium (800,000 Daltons) and high (1,900,000 Daltons) molecular weight HA, the protection against cartilage degeneration in rabbits was significantly greater with the high molecular weight HA two weeks following knee surgery [21]. Their results showed that high molecular weight HA is clinically efficacious in the treatment of incipient osteoarthritis. Hsieh et al (2008) have also shown the benefits of using high molecular weight HA for the treatment of osteoarthritis [26]. In a comparison of high vs. low molecular weight HA, their results showed effective protection for articular cartilage with the high molecular weight HA. This was evidenced by inhibition of MMP-2, MMP-9, u-PA, and PAI-1 expression. Finally, Huang et al (2010) revealed that high molecular weight HA (6,000,000 Daltons) was more effective in downregulating proinflammatory cytokines such as interleukin-1β and tumor necrosis factor-α than lower molecular weight HA (500,000 Daltons to 730,000 Daltons) [27]. The above studies all reaffirm the benefits of higher molecular weight HA, including high viscoelasticity [14, 23], protection of corneal [24, 25] and conjunctival epithelium [24], protection of articular cartilage and anti-inflammation [24, 27].

CONCLUSION

In the characterization of HA raw materials, we have studied and compared hyaluronan data generated by size exclusion chromatography both with triple detection and with multi-angle light scattering detection. Both SEC-TD and SEC-MALS prove to be effective tools to characterize molecular weights and solution properties of HA raw materials. There is good agreement between the two characterization techniques for both the weight-average and number-average molecular weights. The mean M_w values are within 5% and the mean M_n values are within 12% of each other for each of the three lots of HA. The absolute average molecular radii of gyration (R_g) were determined by SEC-MALS. The R_g' values calculated by SEC-TD are slightly smaller than the R_g values determined by SEC-MALS. While SEC-MALS determines the conformation values of HA, SEC-TD is valuable for calculating the intrinsic viscosity and intrinsic viscosity distribution values of HA. All three HA raw materials were determined to have a random coil conformation in the test solvent determined by both techniques. This study has demonstrated that both SEC-TD and SEC-MALS are suitable for characterizing molecular weights, molecular weight distributions, molecular sizes, and solution properties of HA. The knowledge of this detailed molecular information of HA raw materials and potential interactions is tremendously advantageous for process control and predicting product performance in the ophthalmic and personal health care industries.

Aqueous SEC-TD was employed to determine the concentrations of HA and to evaluate the molecular weights and molecular weight distributions of HA present in marketed ophthalmic products. The enzymatic treatment process gives us the higher confidence in identifying and quantifying the HA in various commercially available ophthalmic products. A very broad range of HA molecular weights and concentrations were found in the eleven currently marketed ophthalmic products characterized in this study, with the HA found in Biotrue™ MPS having the highest molecular weight ($p<0.05$). The molecular weight of HA used in ophthalmic products may have impact on clinical and medical performance. More research may be warranted to better understand the clinical and biological implications of the molecular weights of HA in ophthalmic products.

REFERENCES

[1]   Kogan G, Šoltés L, Stern R, Gemeiner P. Hyaluronic acid: a natural biopolymer with a broad range of biomedical and industrial applications. *Biotechnol. Lett.* 2007; 29: 17-25.

[2]   Ali M, Byrne M. Controlled release of high molecular weight hyaluronic acid from molecularly imprinted hydrogel contact lenses. *Pharm. Res.* 2009; 26: 714-726.

[3]   Rah, MJ. A review of hyaluronan and its ophthalmic applications. *Optometry* 2011; 82: 38-43.

[4]   Gomes JAP, Amankwah R, Powell-Richards A, Dua HS. Sodium hyaluronate (hyaluronic acid) promotes migration of human corneal epithelial cells in vitro. *Br. J. Ophthalmol.* 2004;88:821-25.

[5]   Inoue M, Katakami C. The effect of hyaluronic acid on corneal epithelial cell proliferation. *Invest. Ophthalmol. Vis. Sci.* 1993;34:2313-5.

[6]   Nishida T, Nakamura M, Mishima H, Otori T. Hyaluronan stimulates corneal epithelial migration. *Exp. Eye Res.* 1991;53:753-8.

[7]   Fagnola M, Pagani MP, Maffioletti S, Tavazzi S, Papagni A. Hyaluronic acid in hydrophilic contract lenses: Spectroscopic investigation of the content and release in solution. *Cont. Lens Anterior Eye* 2009; 32: 108-112.

[8]   Presti D, Scott JE. Hyaluronan-mediated protective effect against cell damage caused by enzymatically produced hydroxyl (OH·) radicals is dependent on Hyaluronan molecular mass. *Cell Biochem. Funct.* 1994;12:281-88.

[9]   Scott JE. Extracellular matrix, supramolecular organisation and shape. *J. Anat.* 1995;187(Pt2):259-69.

[10]   Szczotka-Flynn LB. Chemical properties of contact lens rewetter. CL *Spectrum* 2006;4.

[11]   Nakamura M, Hikida M, Nakano T, Ito S, Hamano T, Kinoshita S. Characterization of water retentive properties of hyaluronan. *Cornea* 1993;12:433-6.

[12]   Van Beek M, Jones L, Sheardown H. Hyaluronic acid containing hydrogels for the reduction of protein adsorption. *Biomaterials* 2008; 29: 780-789.

[13]   Abbott Medical Optics. http://www.amo-inc.com/products/cataract/ovds 2010.

[14]   Guillaumie F, Furrer P, Felt-Baeyens O, Fuhlendorff B, Nymand S, Westh P, et al. Comparative studies of various hyaluronic acids

produced by microbial fermentation for potential topical ophthalmic applications. *J. Biomed. Mater. Res. A* 2010; 92: 1421-1430.

[15] Fonn D. Targeting contact lens induced dryness and discomfort: what properties will make lenses more comfortable. *Optom. Vis. Sci.* 2007; 84:279-85.

[16] Šoltés L, Mendichi R, Lath D, Mach M, Bakoš D. Molecular characteristics of some commercial high-molecular-weight hyaluronans. *Biomed. Chromatogr.* 2002; 16: 459-462.

[17] Stern R, Asari A, Sugahara K. Hyaluronan fragments: An information-rich system. *Eur. J. Cell Biol.* 2006; 85: 699-715.

[18] Rychlý J, Šoltés L, Stankovská M, Janigová I, Csomorová K, Sasinková V, et al. Unexplored capabilities of chemiluminescence and thermoanalytical methods in characterization of intact and degraded hyaluronans. *Polym. Degrad. Stabil.* 2006; 91: 3174-3184.

[19] Powell JD, Horton MR. Threat Matrix. *Immunol. Res.* 2005; 32: 207-218.

[20] Lyle DB, Breger JC, Baeva LF, Shallcross JC, Durfor CN, Wang NS, et al. Low molecular weight hyaluronic acid effects on murine macrophage nitric oxide production. *J. Biomed. Mater. Res. A* 2010; 94A: 893-904.

[21] Kikuchi T, Yamada H, Shimmei M. Effect of high molecular weight hyaluronan on cartilage degeneration in a rabbit model of osteoarthritis. *Osteoarthr. Cartilage* 1996; 4: 99-110.

[22] Garg HG, Hales CA (eds) *Chemistry and biology of hyaluronan.* Elsevier, Amsterdam, 2004, p 415, 534, 553.

[23] Asari A. Medical Applications of Hyaluronan. In: Garg HG, Hales CA, eds. *Chemistry and Biology of Hyaluronan.* Amsterdam: Elsevier Science; 2004:457-73.

[24] Pauloin T, Dutot M, Joly F, Warnet JM, Rat P. High molecular weight hyaluronan decreases UVB-induced apoptosis and inflammation in human epithelial corneal cells. *Mol. Vis.* 2009; 15:577-83.

[25] Pauloin T, Dutot M, Warnet JM, Rat P. In vitro modulation of preservative toxicity: high molecular weight hyaluronan decreases apoptosis and oxidative stress induced by benzalkonium chloride. *Eur. J. Pharm. Sci.* 2008;34:263-73.

[26] Hsieh YS, Yang SF, Lue KH, Chu SC, Lu KH. Effects of different molecular weight hyaluronan products on the expression of urokinase plasminogen activator and inhibitor and gelatinases during the early stage of osteoarthritis. *J. Orthop. Res.* 2008;26:475-84.

[27] Huang TL, Hsu HC, Yang KC, Yao CH, Lin FH. Effect of different molecular weight hyaluronans on osteoarthritis-related protein production in fibroblast-like synoviocytes from patients with tibia plateau fracture. *J. Trauma* 2010;68:146-52.

[28] Moon MH, Flow field-flow fractionation and multiangle light scattering for ultrahigh molecular weight sodium hyaluronate characterization. *J. Sep. Sci.* 2010; 33: 3519-3529.

[29] Kang DY, Kim WS, Heo IS, Park YH, Lee S. Extraction of hyaluronic acid (HA) from rooster comb and characterization using flow field-flow fractionation (FIFFF) coupled with multiangle light scattering (MALS). *J. Sep. Sci.* 2010; 33: 3530-3536.

[30] Šoltés L, Valachová K, Mendichi R, Kogan G, Arnhold J, Gemeiner P. Solution properties of high-molar-mass hyaluronans: the biopolymer degradation by ascorbate. *Carbohyd. Res.* 2007; 342: 1071-1077.

[31] Liu XM, Gao W, Maziarz EP, Salamone JC, Duex J, Xia E. Detailed characterization of cationic hydroxyethylcellulose derivates using aqueous size exclusion chromatography with on-line triple detection. *J. Chromatogr. A* 2006; 1104: 145-153.

[32] Waters J, Leiske D. Characterization of hyaluronic acid with on-line differential viscometry, multiangle light scattering, and differential refractometry. *LC GC N Am.* 2005; 23: 302-310.

[33] Hokputsa S, Kornelia J, Alexander C, Harding S. A comparison of molecular mass determination of hyaluronic acid using SEC/MALLS and sedimentation equilibrium. *Eur. Biophys. J.* 2003; 32: 450-496.

[34] Viscotek Users Manual and info@viscotek.com, Houston, TX. 2004.

[35] Robbins CR. *Chemical and Physical Behavior of Human Hair*; Van Nostrand Reinhold Company; New York, N.Y. 1979.

[36] Cirrus Multi Detector Software User Guide Revision 3.0. 2007; 13-7, 8, 10.

[37] Zimm BH. The scattering of light and the radial distribution function of high polymer solutions. *J. Chem. Phys.* 1948; 16: 1093-1099.

[38] Gao W, Liu XM, Gross RA. Determination of molar mass and solution properties of cationic hydroxyethyl cellulose derivatives by multi-angle laser light scattering with simultaneous refractive index detection. *Polymer International* 2009; 58(10): 1115-1119.

[39] Cowman MK, Matsuoka S. Experimental approach to hyaluronan structure. *Carbohydrate Res.* 2005; 340(5): 791-809.

[40] Hardingham T. Solution properties of hyaluronan. *In Chemistry and Biology of Hyaluronan*. Eds. H.G. Garg and C.A. Hales. 2004 Elsevier. 1-19.

[41] Terao K, Mays JW. On-line measurement of molecular weight and radius of gyration of polystyrene in a good solvent and in a theta solvent measured with a two-angle light scattering detector. *Eur. Polym. J.* 2004; 40: 1623-1627.

[42] Wyatt PJ. Light scattering and the absolute characterization of macromolecules. *Anal. Chim. Acta* 1993; 272: 1-40.

[43] Jackson C, Chen Y-J, Mays JW. Size exclusion chromatography with multiple detectors: solution properties of linear chains of varying flexibility in tetrahydrofuran. *J. Appl. Polym. Sci.* 1996; 61: 865-874.

[44] Mendichi R., Schieroni AG. Fractionation and characterization of ultra-high molar mass hyaluronan: 2. On-line size exclusion chromatography methods. *Polymer* 2002; 43: 6115-6121.

[45] Mendichi R, Šoltés L, Schieroni AG. Evaluation of Radius of Gyration and Intrinsic Viscosity Molar Mass Dependence and Stiffness of Hyaluronan. *Biomacromolecules* 2003; 4: 1805-1810.

[46] Cowman MK, Mendichi R. Methods for determination of hyaluronan molecular weight. In Chemistry and Biology of Hyaluronan. Eds. H. G. Garg and C. A. Hales. 2004 Elsevier. 41-69.

[47] Podzimek S, Hermannova M, Bilerova H, Bezakova Z, Velebny V. Solution Properties of Hyaluronic Acid and Compariosn of SEC-MALS-VIS Data with Off-Line Capillary Viscometry. *J. Appl. Polym. Sci.* 2010; 116: 3013-3020.

[48] Hokputsa S, Jumel K, Alexander C, Harding SE. Hydrodynamic characterization of chemically degraded hyaluronic acid. Carbohydrate *Polymers* 2003; 52: 111-117.

[49] Walkenhorst, R. Determination of Polymer Structure by Gel Permeation Chromatography. *LC GC Eur.* 2001; 11: 2-4.

[50] Necas J, Bartosikova L, Brauner P, Kolar J. Hyaluronic acid (hyaluronan): a review. *Veterinarni Medicina* 2008; 53(8): 397-411.

[51] Harmon PS, Maziarz EP, Liu XM. Detailed characterization of hyaluronan using aqueous size exclusion chromatography with triple detection and multiangle light scattering detection. *J. Biomed. Mater. Res. B Appl. Biomater.* 2012;100B:1955-60.

[52] Liu XM, Heiler DJ, Menzel T, Brongo A, Burke SE, Cummins K. Sterile Hyaluronic Acid Solutions. US8283463 B2: Oct 9, 2012.

Chapter 6

THE ROLE AND USE OF HYALURONAN IN REPRODUCTIVE MEDICINE

*Barbara Pregl Breznik, Borut Kovačič
and Veljko Vlaisavljević*[*]
Department for Reproductive Medicine and Gyneacologic Endocrinology,
University Medical Center, Maribor, Slovenia

ABSTRACT

This chapter presents the usefulness of hyaluronan in reproductive medicine.

Hyaluronan is a major constituent of the extracellular matrix of the cumulus cells in the cumulus-oocyte complex and may play a critical role in the selection of functionally competent spermatozoa during *in vivo* or *in vitro* fertilization. Hyaluronan can serve as the selective marker during intracytosplasmic sperm injection (ICSI) to select and inject the most optimal spermatozoon. The technique of ICSI requires the immobilization of the spermatozoa. Polyvinylpyrrolidone (PVP) routinely used during ICSI, facilitates handling of spermatozoa. PVP is an artificial polymer, which has been regarded as chemically inert, although adverse effects as a result of its use have been reported. Viscous solution of hyaluronan can be used as an alternative to toxic PVP. Moreover, hyaluronan is also a major glucosaminoglycan in the uterine fluid. It has been shown to increase cell-cell adhesion and may improve embryo apposition and

---

[*] Corresponding author: prof. Veljko Vlaisavljević, PhD - vlai@ukc-mb.si.

attachment. Some studies suggest that the use of hyaluronan containing embryo transfer medium can improve the implantation.

INTRODUCTION

The development of the ICSI technique introduced a new approach to treating male infertility [1]. Almost 20 years later more and more studies point to the unknown side effects of the mentioned method. In the ICSI technique the spermatozoa are artificially selected, not allowing the natural selection to take place among them.

Wrong selection of a spermatozoon could cause the injection of the spermatozoon with fragmented DNA, which could later result in the decreased level of the fertilization, affect the quality of the embryo and the embryo's implantation ability [2-9] or increase the possibility of spontaneous abortion [9-14].

Studies also emphasize the higher share of children conceived by the ICSI method with congenital defects compared to children conceived by the classic IVF (in vitro fertilization) method or naturally conceived children [15-17]. The right selection of the spermatozoon is crucial for the successful outcome following the ICSI technique, therefore reproductive medicine is searching for methods which would enable a more reliable selection of spermatozoa for intracytosplasmic injection.

Hyaluronan, which is the integral part of extracellular matrix of cumulus cells surrounding the oocyte, represents such an option. Namely, only spermatozoa without DNA defragmentation, cytoplasmic retention and proper morphology has the hyaluron-binding ability [18-20]. This ability is used in reproductive medicine to select the spermatozoon for intracytoplasmic injection, for evaluating the fertilization ability of spermatozoa or explaining male infertility.

In addition to the artificial selection of the spermatozoon in the ICSI method, the use of polyvinylpyrrolidone (PVP) for the necessary immobilization of spermatozoa during the procedure also has a negative effect on the procedure's outcome. Hyaluronan as a viscous solution represents a successful substitute for toxic PVP [21, 22].

Hyaluronan is also present in cervical mucus. Using hyaluronan in the embryo transfer medium can significantly affect the embryo's successful implantation [23-26]. In the following section we present the mentioned options of using hyaluronan in reproductive medicine.

THE DEVELOPMENT OF HYALURONIC ACID IN THE EXTRACELLULAR MATRIX OF THE CUMULUS CELLS AND ITS ROLE

The human oocyte is surrounded by several layers of cumulus cells (Figure 1) which provide the necessary nourishment and hormones during its development and growth. On the luteinizing hormone, which is a signal for oocyte maturation and ovulation, numerous changes are triggered. The cumulus cells begin to build the extracellular matrix, which mostly consists of hyaluronic acid [27].

Hyaluronic acid or hyaluronan is a glycosaminoglycan, a linear polymer made of repeating disaccharide, D-glucuronic acid units (1-β-3) and N-acetyl-D-glucosamine (1-β-4). Hyaluronan is synthesised by means of integral membrane proteins called hyaluronan synthases found in three forms: HAS1, HAS2 and HAS3. Enzymes build a chain by attaching the glucuronic acid and N-acetyl glucosamine intermittently on the developing polysaccharide.

Synthesised hyaluronic acid transfers, with the help of transfer protein, through the cell membrane into the extracellular space and is thus the integral part of the extracellular matrix of the cumulus cells surrounding the oocyte [28, 29]. The molecules of the hyaluronan are in the extracellular matrix of the cumulus cells cross-connected with proteins of the extracellular matrix and proteoglycans. Such extracellular matrix of the cumulus cells develops not sooner than and only before the ovulation. Hyaluronan incorporates into the gap junctions of cumulus cells.

By doing so, it loosens the junctions between cumulus cells as well as their attachment to the oocyte [30] and enables the spermatozoon's passage through the cumulus cells to the oocyte [31].

A high concentration of hyaluronan in the follicular fluid is linked with a successfully fertilized oocyte [31].

SPERMATOGENESIS AND ACQUIRING THE HYALURONIC ACID BINDING ABILITY

The course of spermatogenesis has three stages: proliferation phase, meiotic phase, and differentiation phase (spermiogenesis). It takes place in seminiferous tubules of the testes and right at the basal lamina lays a layer of spermatogonias (with diploid number of chromosomes).

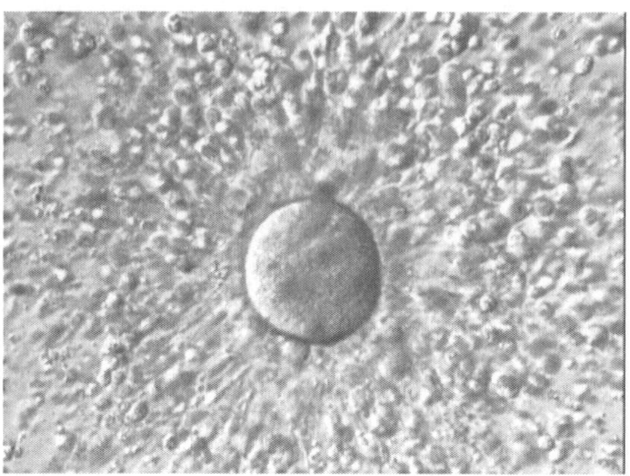

Figure 1. Oocyte surrounded by several layers of cumulus cells (oocyte cumulus complex).

Spermatogonias divide mitotically (proliferation phase). Some of the cells which are the result of spermatogonial cell division develop into primary spermatocytes. The primary spermatocytes enter into meiotic division, resulting in haploid round spermatids (meiosis). During spermiogenesis the spermatozoa develops from the round spermatid. During this process histones are eliminated and replaced by protamines.

Moreover, a part of the nuclear membranes, half of mitochondria and a larger part of cytoplasm are discharged, an acrosome develops, and receptors are built into the membrane for binding to zona pellucida and binding to hyaluronan.

As the receptors for binding to zona pellucida and binding to hyaluronan have the same origin and are built into the membrane simultaneously, the two characteristics are closely connected [32].

THE USE OF HYALURONAN AS A SELECTIVE MARKER FOR SELECTING THE OPTIMAL SPERMATOZOON FOR INTRACYTOSPLASMIC INJECTION

The Importance of Selecting the Most Optimal Spermatozoon

In procedures for treating infertility with IVF we use two methods for fertilizing the oocytes; the intracytoplasmic sperm injection (ICSI) (Figure 2) and the classic *in vitro* fertilization (IVF) method.

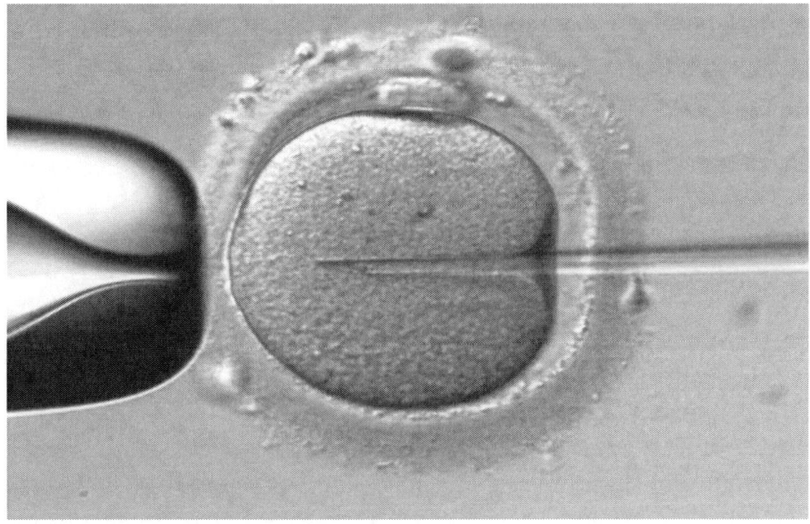

Figure 2. Intracytosplasmic sperm injection (ICSI).

In the latter, we only bring the spermatozoa closer to the oocyte and they must come into the oocyte by himself. In the ICSI technique we physiccally deposit the spermatozoon (with the injection needle) into the cytoplasm of the oocyte.

Compared to IVF, the ICSI technique provides a higher share of fertilized oocytes and lower probability of total fertilization failure [33-35]. But on the other hand, in the IVF method we allow the natural selection between spermtozoa to take place, whereas in the ICSI technique, the selection is performed by an embryologist, who first provides a subjective evaluation of motility and morphology of the spermatozoon and then selects the most appropriate one. The success of the ICSI technique mostly depends on the selection of the spermatozoon injected.

The spermatozoon´s hyaluron-binding ability is one of the biomarkers based on which we select the appropriate spermatozoon. This selection is based on the fact that the spermatozoon requires this ability to recognize and bind to hyaluronan also *in vivo* for successful fertilization [36]. Only mature spermatozoon binds to hyaluronan (through receptors on its surface), with normal chromosomal status, without retention of cytoplasm, the remaining unreplaced histones (residual histones) or apoptotic markers [19, 37] and without fragmented DNA [2, 20]. The most important characteristic of spermatozoon injected into oocyte is the integrity of spermatozoon's DNA, which is not visible to the embryologist who selects the spermatozoon on the basis of motility and morphology and is not reflected in the spermatozoon's altered morphology.

Avendano et al. [38] proved that even seemingly morphologically normal spermatozoa can have fragmented DNA. The consequences of inserting a spermatozoon with fragmented DNA are usually seen no earlier than after three days of embryo cultivation.

A high share of spermatozoa with fragmented DNA is related to recurring unsuccessful IVF procedures [2-9] and a higher share of spontaneous abortions [9-14]. Fertilizing the oocyte with a spermatozoon with fragmented DNA, if it happens at all, affects the embryo's vitality in later stages of development (prior to implantation and after it).

The mentioned effect is especially noticeable in embryos developing after the applied ICSI method where on account of the artificial selection of sperm the probability of injecting a spermatozoon with fragmented DNA is higher than in the IVF method [39]. Studies point to a higher share of children conceived by the use of the IVF method or naturally conceived children[15-17]. Zona pellucida has the ability to select among optimal and non-optimal spermatozoa *in vivo* [40, 41]. The spermatozoon without the hyaluronic-binding ability could never fertilize an oocyte *in vivo* as it does not have the ability to bind to zona pellucida which is unavoidable in the fertilization process (Figure 3). Thus the spermatozoon's ability to bind to hyaluronan defines its fertilizing ability and the selection for injecting the spermatozoon into the oocyte during the ICSI procedure becomes very similar to *in vivo* selection. Thus two laboratory techniques have been developed. The first one enables the selection of sperm for intracytoplasmic injection and the second one assesses the spermatozoa's fertilizing ability (hyaluronan-binding test).

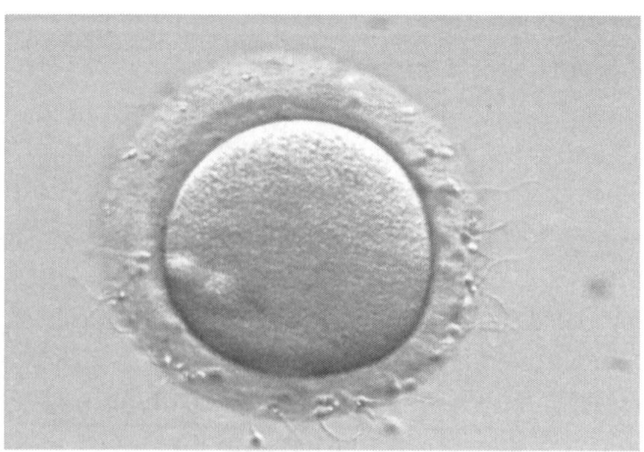

Figure 3. Spermatozoa binding to zona pellucida.

Selection of Spermatozoon for Injection into the Oocyte during the ICSI Procedure on the Basis of Spermatozoon's Hyaluronan-Binding Ability

In this method we use a Petri dish with hyaluronic acid bound to it. This method enables us during the ICSI procedure to immobilize the mature spermatozoa binding to hyaluronan [42]. The share of aneuploid spermatozoa binding to hyaluronan on the Petri dish is 4-6 fold lower in comparison to the fraction obtained after a swim-up method, widely used for sperm preparation [18]. The use of hyaluronan for selecting spermatozoa intended for the ICSI technique increases the level of oocyte fertilization [43], embryo quality [44], and pregnancy rate as well as lowers the occurrence of spontaneous abortions in comparison to patients where the selection of spermatozoa for intracytoplasmic injection was performed on the basis of the embryologist's assessment [43].

Hyaluronan Binding Test

The spermatozoa's ability to bind to hyaluronic acid is tested on special glass slides coated with hyaluronic acid (Figure 4). The test is available on the market under the name Hyluronan binding assay (HBA®).

To perform the test, 10 microlitres of a sample are required and the result is visible after 10 minutes. Bound spermatozoa are differentiated from unbound spermatozoa by their beating tails with heads that make no progressive movement. Non-binding motile spermatozoa swim about freely.

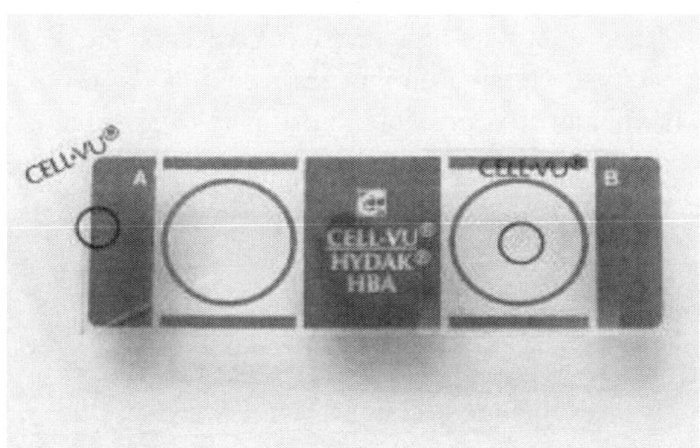

Figure 4. Glass slides for performing the hyaluronan binding test [45].

The test result is related to the spermatozoa's fertilizing ability and fertilization rate following the IVF method. The hyaluronan binding ability correlates with the percentage of morphologically optimal spermatozoa [36].

The Use of Hyaluronan Binding Test for Establishing the Spermatozoa's Fertilizing Ability and Differentiating among Samples Appropriate for the IVF or ICSI Method

The criteria for evaluating the quality of spermatozoa, is a classic semen analysis, which evaluates the concentration, motility and percentage of morphologically optimal spermatozoa. Based upon the obtained values we decide for one of the two methods for fertilizing the oocytes.

The ICSI technique is indicated only in cases with severe forms of male infertility (heavily decreased sperm concentration or motility, globozoospermia, complete absence of spermatozoa in the ejaculate, immunological cause of infertility) or in special cases when it is clear from previous IVF attempts that the percentage of fertilized oocytes after the IVF method was extremely low. The percentage of couples with one of the mentioned causes of infertility is around 30 % [47]. Despite the mentioned indications for employing the ICSI technique, this method is used in IVF procedures across Europe in almost 70% of all assisted reproduction techniques (ART) cycles [48]. The reason lies in the fact that despite normal values resulting from the classic semen analysis, the male infertility factor cannot be excluded [49-54]. In recent years a need for additional testing was created to obtain a more reliable assessment of the fertilizing ability and functional maturity of spermatozoa.

In the following section we will describe the study conducted at the Department of Reproductive Medicine and Gynaecologic Endocrinology of the University Medical Centre Maribor, Slovenia. The purpose of this study was to establish whether it is possible to estimate of spermatozoa's fertilizing ability based on the hyaluronan-binding test and thus differentiate among semen samples suitable for one of the two insemination methods.

Methods

Group of Patients

In this study we included couples who were in the first or second procedure of treating infertility with ART and where we collected at least 6 oocytes after ovarian stimulation with an ultrasound-guided ovarian puncture.

Women were younger than 37 years and the cause of infertility was not endometriosis, polycystic ovaries (PCO) or polycystic ovary syndrome (PCOS). The couples with an identified severe form of male infertility were not included in the study. A half of the oocytes were fertilized by the ICSI technique and the other half with the IVF method.

Analysis on a Sperm Sample

Semen samples were collected on the day of the ultrasound-guided ovarian puncture by masturbation after two to five days of sexual abstinence.

The semen was prepared with the density gradient technique (Puresperm, Nidacon, Švedska) and afterwards with the swim-up method. The supernatant with progressively motile spermatozoa was used for insemination of oocytes and performance of hyaluronan-binding test.

Insemination of Oocytes and Cultivation of Embryos

Half of the oocytes were inseminated with the ICSI technique and the other half with the classic IVF method. The inseminated oocytes were incubated for 18-20 hours at 37 °C, 95 % humidity, 6 % CO_2 and 5 % O_2.

The following day the zygotes were moved to a sub-cultivation where we separated the correctly fertilized from incorrectly fertilized or unfertilized oocytes. We identified oocytes as correctly fertilized if they had after 18-20 hours visible two pronuclei (2PN) in the cytoplasm of the oocyte.

Results

133 couples successfully completed the study. The hyaluronan-binding test in 133 prepared semen samples reached values from 15% to 100% (with the average value 91.3 ± 10.9) (Figure 5).

The Effect of the Share of Spermatozoa with Hyaluronan-Binding Ability on the Fertilization of Oocytes After the IVF Method

We have established that the percentage of spermatozoa with hyaluronan-binding ability is linked with the percentage of fertilized oocytes following the IVF method (R=0.321, P=0.000) (Figure 6).

After the ICSI method we did not establish that the percentage of spermatozoa with hyaluronon-binding ability would affect the fertilization of oocytes (R=0.114, P=0.190).

The reason probably lies in the fact that the ICSI method bypasses the natural course of oocyte fertilization, more specifically, the entrance of the spermatozoon into the oocyte.

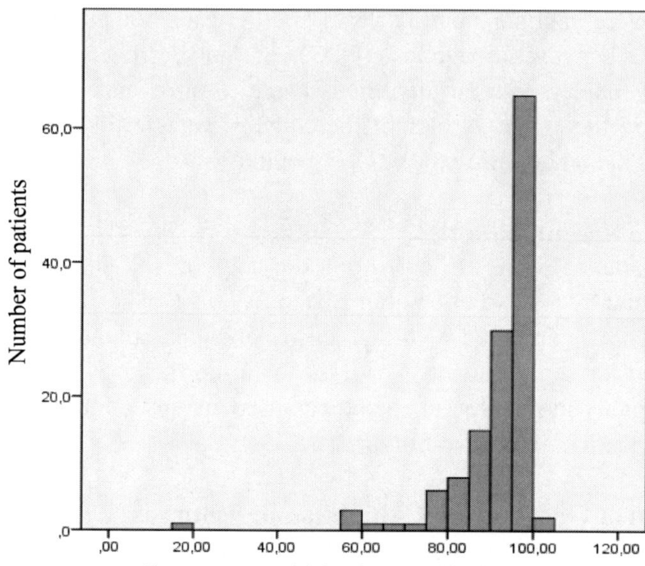

Figure 5. Distribution of results of hyaluronan-binding test on a prepared semen sample in the study male group (N = 133).

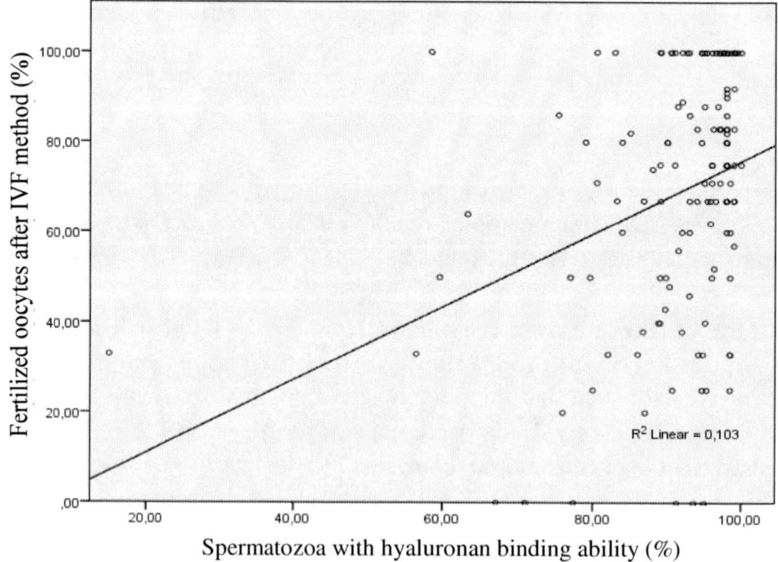

Figure 6. Percentage of fertilized oocytes following the IVF method in 133 couples regarding the percentage of spermatozoa bound to hyaluronan in a prepared semen sample.

This way the spermatozoon does not acquire the hyaluronan-binding ability as it is mechanically injected into the cytoplasm of the oocyte. Our results are in accordance with certain other studies [2, 36, 44, 55].

A Comparison of Groups with the Fertilization Rate of Oocytes Following the IVF Method of Less Than 50% and More Than 50%

Couples included in the study were divided into two groups. Group 1 was represented by couples with a fertilization rate of less than 50% following the IVF method and Group 2 was presented by couples with a fertilization rate of more than 50% following the IVF method. The groups did not differ regarding the characteristics which could significantly affect the fertilization of oocytes, i.e. mean age of women, number of oocytes collected with ultrasound-guided ovarian puncture, women's BMI or quantity of applied gonadotrophins for ovarian stimulation.

Table 1. Comparison of groups with the fertilization rate following the IVF method less (Group 1) and more than 50% (Group 2) in characteristics of the groups and characteristics of the semen

	Group 1 Fertilization rate < 50 % (N = 29)	Group 2 Fertilization rate ≥ 50 % (N = 104)	P value
Group characteristics			
Woman's age	31.5 ± 0.6	32.3 ± 0.3	NS
Woman's BMI	22.3 ± 0.6	22.1 ± 0.3	NS
Number of used gonadotrophin ampoules	23.5 ± 1.3	23.8 ± 0.6	NS
Number of collected oocytes	12.5 ± 1.4	13.4 ± 0.7	NS
Native semen sample			
Concentration ($*10^6$/ml)	44.1 ± 8.5	64.2 ± 4.7	0.045
Progressive motility (%)	30 ± 2.1	35.1 ± 1.3	NS
Morphologically normal shape (%)	5.2 ± 0.7	6.4 ± 0.4	NS
Prepared semen sample			
Hyaluronan-binding ability (%)	85.1 ± 3.1	93 ± 0.8	0.019

Values are expressed as the mean value ± *standard error mean, SEM)*.

P value = measure for statistical significance, $P < 0,05$ statistical significance.

NS = difference not statistically significant.

There were also no differences between the groups in the quality of the native semen sample (motility and percentage of morphologically normal shapes). The only slight difference was in the sperm concentration in native samples (P = 0.045).

The groups differed in statistical significance regarding the percentage of spermatozoa with hyaluronan binding ability (P = 0.019) defined in a prepared semen samples used for later insemination of oocytes (Table 1).

By using the a receiver operating characteristic (ROC) curve we set the limit value for the percentage of spermatozoa with hyaluronan-binding ability on the basis of the achieved fertilization rate of more or less than 50% following the IVF method. This limit value was set to 90.85 % with 75 % sensitivity and 55.2 % specificity. Based on this we are able to separate semen samples suitable for the IVF method or ICSI method because we can anticipate the percentage of fertilized oocytes following the IVF method. The curve is statistically significant ($P = 0.000$) with the area below the curve (AUC^{ROC}) 0.721 (0.618–0.823) (95 % CI) (Figure 7).

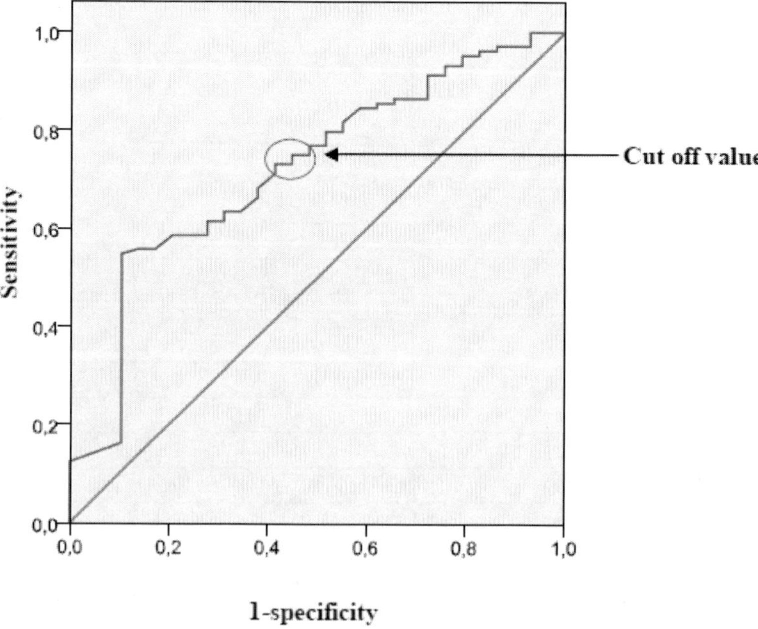

Figure 7. ROC curve representing the percentage of spermatozoa with hyaluronan-binding ability regarding the percentage of fertilized oocytes following the IVF method lower or higher than 50%.

Conclusion

We established in our study that the hyaluronan-binding ability analysed on a semen sample has s significant impact on the percentage of fertilized oocytes following the IVF method. Based on the result of the mentioned test we can estimate with certainty the fertilizing ability of spermatozoa in the IVF method and successfully differentiate the semen samples suitable for the IVF method from those less appropriate on account of the expected low share of fertilized oocytes (less than 50%).

WHY CAN THE HYALURONAN-BINDING ABILITY BE AFFECTED (CAUSES)?

Reactive Oxygen Species

Reactive oxygen species (ROS) are highly reactive free radicals. They are generated as undesirable side products of the oxidative energy metabolism.

From the biological perspective the following three compounds are especially important: hydrogen peroxide (H_2O_2), superoxide or superoxide anion radical ($O_2^{\cdot-}$) and hydroxyl radical ($OH^{\cdot-}$).

When the level of their concentration exceeds the limits tolerable and manageable for the cell, oxidative stress occurs and causes cell damage [56]. Free oxygen radicals with lipid peroxidation (oxidation of unsaturated fatty acids) damage the lipids in the cell membrane.

Spermatozoon's membrane is due its composition especially susceptible to ROS-induced damage. 50% of the spermatozoon's membrane is represented by the docosahexaenoic acid (DHA). DHA is a 22-carbon chain with six double bonds (polyunsaturated fatty acids). Oxidation of these fatty acids leads to a loss of fatty acids from the cell membrane structure, which causes instability and dysfunction of the cell membrane [57].

Therefore the membrane looses the fluidity, ability to maintain ion gradient and successful course of transport processes as well as the position of receptors in the membrane [58].

The spermatozoon is under different conditions exposed to excessive and unmanageable function of reactive oxygen species, which differ according to source and place of function:

* The spermatozoa travel after the completed spermatogenesis from the seminiferous tubules of the testes into the epididymis and remain there for days [59]. Among them are also immature or even dead spermatozoa which are the source of ROS.

* Banks et al. [60] concluded in his study that exposure of testes to a high temperature (42 °C) evokes an increased ROS concentration in the epididymis as a reaction to higher temperature. Epididymis at higher temperature is no longer an appropriate environment for spermatozoa, which is reflected in the increased share of immotile and dead spermatozoa in the epididymis [61, 62], which are afterwards an additional source of ROS.

* Leucocytospermia (presence of leukocytes in the semen, their concentration higher than 1 million/ml): peroxidase-positive leukocytes (poly-morphonuclear leukocytes and macrofagues stem from the prostate and seminal vesicles [63] are the source of ROS [64, 65]

* smoking: nicotine has been linked with the increased production of ROS [66-68]

* alcohol: ROS are generated in metabolic pathways for the conversion of ethanol [69]

It is clear from the above stated that a spermatozoon is inevitably exposed to the function of ROS which have a different source and the duration of the exposure varies. However, the consequence of their action is the dysfunction of the membrane, affected structure and loss of receptors for binding to hyaluronan.

Irregularities in the Course of Spermiogenesis

During spermiogenesis the spermatozoa develop from the round spermatids. During this process histones are discarded and replaced by protamines, a part of the nuclear, half of the mitochondria and most of the cytoplasm are discharged, an acrosome develops, receptors are built into the membrane for binding to zona pellucida and binding to hyaluronan [32]. If a spermatozoon from the semen is exposed to hyaluronic acid, we can notice that certain spermatozoa successfully bind to hyaluronic acid (have acquired during spermiogenesis the receptors for binding to hyaluronan) and some not. Analysing the characteristics of spermatozoa with binding ability led to the conclusion that these are in fact spermatozoa which discharged excessive cytoplasm during spermiogenesis, successfully set their chromosomal statue during meiosis and do not have fragmented DNA which is known to largely develop during spermatogenesis [19, 37]. This basically means that in order to successfully develop the hyaluronan-binding ability it is necessary and important that all stages of spermatogenesis are progressing normally. The spermatozoa which have successfully completed the spermiogenesis and successfully bind to hyaluronan also share certain other characteristics, which explains the cause and effect link between the mentioned characteristics.

Huszar et al. established in one of his studies that the quantity of creatine kinases, which is normally a measure for the quantity of remaining cytoplasm, correlates with the fertilizing ability of spermatozoa *in vivo* or male fertility where a successful transformation of spermatozoa during spermiogenesis, including the acquisition of hyaluronan-binding ability, of key importance [70].

The quantity of HspA2p protein is also related to the fertilizing ability of spermatozoa. HspA2 is a part of sinaptonemal complex included into the intracellular transport of enzymes for DNA repair mechanisms, explaining why only those spermatozoa which have a normal chromosomal status and no fragmented DNA, have the hyaluronan-binding ability [71].

USE OF HYALURONAN INSTEAD OF POLYVINYLPYRROLIDONE

In the ICSI method it is required to immobilize the spermatozoon for a successful selection and intracytoplasmic injection. The most commonly used medium for this process is polyvinylpyrrolidone (PVP). It is a large (molecule weight 360 kDa) water-soluble, highly viscous polymer made from the monomer N-vinylpyrrolidone [72].

During the process of selection and injection, we expose the spermatozoon to PVP and at injection insert a small amount of it together with the spermatozoon into the cytoplasm of the oocyte [73]. Despite the frequent use of PVP, studies have shown that exposure of spermatozoa to PVP cause irreparable damage to the spermatozoa. PVP has a negative effect on the important structures of spermatozoon, such as acrosome, mitochondria, and plasma membrane [74]. After intracytoplasmic injection, PVP interrupts the characteristic calcium oscillations, which are important for the fertilization process and even interrupts the decondensation of sperm chromatin [75].

Consequently, it affects the percentage of fertilized oocytes, embryo development [76-78] and embryo chromosome irregularities [79].

As PVP is a large polimer, it cannot be discharged from the oocytes and is not decomposable with cell enzymes and thus remains in the oocyte or embryo longer [80].

On account of the mentioned characteristics of PVP, an alternative and more physiological medium is increasingly used. Like PVP, this medium has to be highly viscous to successfully slow the spermatozoa, their attachment to the bottom of the dish and has no influence on the fertilization of the oocyte or further embryo development.

Hyaluronan is in addition to the stated characteristics, naturally present in the human organism and as part of the extracellular matrix of the cumulus cells directly participates in the fertilization process of the oocyte.

Unlike PVP, is decomposes inside the cell into monosaccharides in the usual biochemical manner and therefore represent a more physiological alternative to the normally used PVP.

Several studies confirm the successful transition from PVP to hyaluronan and successful outcome of the ICSI procedure [21, 22].

HYALURONAN IN EMBRYOTRANSFER MEDIUM

Hyaluronan can be found besides in cumulus cells also in cervical mucus, fallopian tube, uterus and follicular fluid [81]. The amount of hyaluronan in the endometrium changes during the menstrual cycle, an increased amount of hyaluronan in the endometrium is noticeable during the receptive phase of the endometrium [82] and the receptors for binding to hyaluronan on the surface of the human embryo through the entire pre-implantation phase [83]. They are the so-called CD44 receptors from the family of cadherins, by means of which the embryo is able to bind to hyaluronan. All this indicates the important role of hyaluronan in the pre-implantation process [84] where hyaluronan stimulates important intercellular interactions during the initial stages of the adhesion of the blastocyst to the endometrium [85, 86]. Several studies have been performed with the purpose to establish the effect of hyaluronan (presence in the embryotransfer medium, EmbryoGlue (Vitrolife, US)) of the success of embryo implantation. The studies compared the implantation rate and percentage of clinically confirmed pregnancies between groups where the embryotransfer was carried out from a medium containing hyaluronan (EmbryoGlue) and from a medium without it. Several studies have confirmed a positive effect of using hyaluronan medium on the implantation of the embryo [24-26, 87].

A positive effect of hyaluronan on the percentage of clinical pregnancies was confirmed by less studies [25, 26, 87, 88], while the remaining could not confirm this effect [24, 89-91].

CONCLUSION

Ubiquitarian hyaluronic acid in the human body represents an important compound with great significance in reproductive medicine. The sperm's hyaluronan-binding ability is unique. By observing this binding we are able to estimate its fertilizing ability, recognize the causes of male infertility and contribute to a successful outcome of treating infertility.

REFERENCES

[1]   Palermo, G. D., Cohen, J., Alikani, M., Adler, A., Rosenwaks, Z. Intra-cytoplasmic sperm injection: a novel treatment for all forms of male factor infertility. *Fertil. Steril.* 1995; 63: 1231-40.

[2]   Nasr-Esfahani, M. H., Razavi, S., Vahdati, A. A., Fathi, F., Tavalaee, M. Evaluation of sperm selection procedure based on hyaluronic acid binding ability on ICSI outcome. *J. Assist. Reprod. Genet.* 2008; 25: 197-203.

[3]   Lopes, S., Sun, J. G., Jurisicova, A., Meriano, J., Casper, R. F. Sperm deoxyribonucleic acid fragmentation is increased in poor-quality semen samples and correlates with failed fertilization in intracytoplasmic sperm injection. *Fertil. Steril.* 1998; 69: 528–32.

[4]   Sun, J. G., Jurisicova, A., Casper, R. F. Detection of deoxyribonucleic acid fragmentation in human sperm: correlation with fertilization in vitro. *Biol. Reprod.* 1997; 56 (3): 602-7.

[5]   Larson, K. L., DeJonge, C. J., Barnes, A. M., Jost, L. K., Evenson, D. P. Sperm chromatin structure assay parameters as predictors of failed pregnancy following assisted reproductive techniques. *Hum. Reprod.* 2000; 15: 1717-22.

[6]   Morris, I. D. Sperm DNK damage and cancer treatment. *Int. J. Androl.* 2002; 25: 255-61.

[7]   Benchaib, M., Braun, V., Lornage, J., Hadj, S., Salle, B., Lejeune, H., Guérin, J. F. Sperm DNA fragmentation decreases the pregnancy rate in an assisted reproductive technique. *Hum. Reprod.* 2003; 18 (5): 1023-8.

[8]   Muriel, L., Garrido, N., Fernandez, J. L., Remohi, J., Pellicer, A., de los Santos, M. J., Meseguer, M. Value of the sperm deoxyribonucleic acid fragmentation level, as measured by the sperm chromatin dispersion test, in the outcome of in vitro fertilization and intracytoplasmic sperm injection. *Fertil. Steril.* 2006; 85 (2): 371-83.

[9]   Cebesoy, F. B., Aydos, K., Unlu, C. Effects of chromatin damage on fertilization ratio and embryo quality post-ICSI. *Arch. Androl.* 2006; 52: 397-402.

[10] Benchaib, M., Lornage, J., Mazoyer, C., Lejeune, H., Salle, B., François Guerin, J. Sperm deoxyribonucleic acid fragmentation as a prognostic indicator of assisted reproductive technology outcome. *Fertil. Steril.* 2007; 87 (1): 93-100.

[11] Borini, A., Tarozzi, N., Bizzaro, D., Bonu, M. A., Fava, L., Flamigni, C., Coticchio, G. Sperm DNA fragmentation: paternal effect on early post-implantation embryo development in ART. *Hum. Reprod.* 2006; 21 (11): 2876-81.

[12] Zini, A., Meriano, J., Kader, K., Jarvi, K., Laskin, C. A., Cadesky, K. Potential adverse effect of sperm DNA damage on embryo quality after ICSI. *Hum. Reprod.* 2005; 20 (12): 3476-80.

[13] Check, J. H., Graziano, V., Cohen, R., Krotec, J., Check, M. L. Effect of an abnormal sperm chromatin structural assay (SCSA) on pregnancy outcome following (IVF) with ICSI in previous IVF failures. *Arch. Androl.* 2005; 51 (2): 121-4.

[14] Carrell, D. T., De Jonge, C., Lamb, D. J. The genetics of male infertility: a field of study whose time is now. *Arch. Androl.* 2006; 52 (4): 269-74.

[15] Bonduelle, M., Wennerholm, U. B., Loft, A., Tarlatzis, B. C., Peters, C., Henriet, S., Mau, C., et al. A multi-centre cohort study of the physical health of 5-year-old children conceived after intracytoplasmic sperm injection, in vitro fertilization and natural conception. *Hum. Reprod.* 2005; 20 (2): 413-9.

[16] Ludwig, M., Katalinic, A. Malformation rate in fetuses and children conceived after ICSI: results of a prospective cohort study. *RBM Online* 2002; 5 (2): 171-8.

[17] Katalinic, A., Rosch, C., Ludwig, M. Pregnancy course and outcome after intracytoplasmic sperm injection: a controlled, prospective cohort study. *Fertil. Steril.* 2004; 81 (6): 1604-16.

[18] Jakab, A., Sakkas, D., Delpiano, E., Cayli, S., Kovanci, E., Ward, D., Ravelli, A., et al. Intracytoplasmic sperm injection: a novel selection method for sperm with normal frequency of chromosomal aneuploidies. *Fertil. Steril.* 2005; 84, 6: 1665-73.

[19] Huszar, G., Celik-Ozenci, C., Cayli, S., Zayaczki, Z., Hansch, E., Vigue, L. Hyaluronic acid binding by human sperm indicates cellular maturity, viability and unreacted acrosomal status. *Fertil. Steril.* 2003; 79: 1616-24.

[20] Parmegiani, L., Cognigni, G. E., Bernardi, S., Torilo, E., Ciampaglia, W., Filicori, M. »Physiologic ICSI«: Hyaluronic acid (HA) favors selection of spermatozoa without DNK fragmentation and with normal nucleus, resulting in improvement of embryo quality. *Fertil. Steril.* 2010; 93 (2): 598-604.

[21] Barak, Y., Menezo, Y., Veiga, A., Elder, K. A physiological replacement for polyvinylpyrrolidone (PVP) in assisted reproductive technology. *Hum. Fertil.* 2001; 4: 99-103.

[22] Balaban, B., Lundın, K., Morrell, J. M., TjellstroĖm, H., Urman, B., Holmes, P. V. An alternative to PVP for slowing sperm prior to ICSI. *Hum. Reprod.* 2003; 18 (9): 1987-9.

[23] Schoolcraft, W., Lane, M., Stevens, J., Gardner, D. K. Increased hyaluronan concentration in the embryo transfer medium results in a significant increase in human embryo implantation rate. *Fertil. Steril.* 2002; 78 (1): 5.

[24] Hazlett, W. D., Meyer, L. R., Nasta, T. E., Mangan, P. A., Karande, V. C. Impact of EmbryoGlue as the embryo transfer medium. *Fertil. Steril.* 2008; 90 (1): 214-6.

[25] Friedler, S., Schachter, M., Strassburger, D., Esther, K., Ron El, R., Raziel, A. A randomized clinical trial comparing recombinant hyaluronan/recombinant albumin versus human tubal fluid for cleavage stage embryo transfer in patients with multiple IVF-embryo transfers: a prospective randomized study. *Hum. Reprod.* 2007; 22 (9): 2444-8.

[26] Urman, B., Yakin, K., Ata, B., Isiklar, A., Balaban, B. Effect of hyaluronan-enriched transfer medium on implantation and pregnancy rates after day 3 and day 5 transfers: a prospective randomized study. *Fertil. Steril.* 2008; 90 (3): 604-12.

[27] Russell, D. L., Salustri, A. Extracellular matrix of the cumulus-oocyte complex. *Semin. Reprod. Med.* 2006; 24(4): 217-27.

[28] Laurent, T. C., Fraser, J. R. Hyaluronan. *FASEB J.* 1992; 6 (7): 2397–2404.

[29] Schulz, T., Schumacher, U., Prehm, P. Hyaluronan export by the ABC transporter MRP5 and its modulation by intracellular cGMP. *J. Biol. Chem.* 2007, 282: 20999-21004.

[30] Salustri, A., Yanagishita, M., Underhill, C. B., Laurent, T. C., Hascall, V. C. Localization and synthesis of hyaluronic acid in the cumulus cells and mural granulosa cells of the preovulatory follicle. *Developmental Biology* 1992; 151: 541-551.

[31] Saito, H., Kaneko, T., Takahashi, T., Kawachiya, S., Saito, T., Hiroi, M. Hyaluronan in follicular fluids and fertilization of oocytes. *Fertil. Steril.* 2000; 74: 1148-1152.

[32] Elder, K., Dale, B. *In-Vitro fertilization*, 3rd edition. Anglija: Cambridge University Press; 2011.

[33] Xi, Q., Zhu, L., Hu, J., Wu, J., Zhang, H. Should few retrieved oocytes be as an indication for intracytoplasmic sperm injection? *Biomed. Biotechnol.*, 2012; 13: 717-22.

[34] Fang, C., Tang, J., Huang, R., Li, L. L., Zhang, M. F., Liang, X. Y. Comparison of IVF outcomes using conventional insemination and ICSI in ovarian cycles in which only one or two oocytes are obtained. *J. Gynecol. Obstet. Biol. Reprod.*, 2012; 6: 1-7.

[35] Gozlan, I., Dor, A., Farber, B., Meirow, D., Feinstein, S., Levron, J. Comparing intracytoplasmic sperm injection and in vitro fertilization in patients with single oocyte retrieval. *Fertil. Steril.* 2007; 87: 515-8.

[36] Ye, H., Huang, G., Gao, Y., Liu De, Y. Relationship between human sperm hyaluronan binding assay and fertilization rate in conventional in vitro fertilization. *Hum. Reprod.* 2006; 21: 1545-50.

[37] Cayli, S., Jakab, A., Ovari, L., Delpiano, E., Celik-Ozenci, C., Sakkas, D., et al. Biochemical markers of sperm function: male fertility and sperm selection for ICSI. *Reprod. Biomed. Online* 2003; 7: 462-8.

[38] Avendano, C., Franchi, A., Taylor, S., Morshedi, M., Bocca, S., Oehninger, S. Fragmentation of DNA in morphologically normal human spermatozoa. *Fertil. Steril.* 2009; 92: 835-48.

[39] Aitken, R. J., Baker, M. A. Oxidative stress and male reproductive biology. *Reprod. Fertil. Dev.* 2004; 16 (5): 581-8.

[40] Menkveld, R., Franken, D. R., Kruger, T. F., Oehninger, S., Hodgen, G. D. Sperm selection capacity of human zona pellucida. *Mol. Reprod. Dev.* 1991; 30 (4): 346-52.

[41] Van Dyk, Q., Lanzendorf, S., Kolm, P., Hodgen, D. J., Mahony, M. C. Incidence of aneuploid spermatozoa from subfertile men: selected with motility versus hemizona bound. *Hum. Reprod.* 2000; 15: 1529-36.

[42] Huszar, G., Jakab, A., Sakkas, D., Ozenci, C. C., Cayli, S., Delpiano, E., Ozkavukcu, S. Fertility testing and ICSI sperm selection by hyaluronic acid binding: clinical and genetic aspects. *Reprod. Biomed. Online* 2007; 14, 5: 650-63.

[43] Worrilow, K. C., Huynh, H. T., Bower, J. B., Anderson, A. R., Schillings, W., Crain, J. L. PICSI vs. ICSI: statistically significant improvement in clinical outcomes in 240 in vitro fertilization (IVF) patients. *Fertil. Steril.* 2007; 88, 1: 37.

[44] Bacer – Kermavner, L., Virant-Klun, I., Vrtačnik-Bokal, E., Tomaževič, T., Meden Vrtovec, H. Sperm selection with hyaluronan and blastocyst development after ICSI. *Zdrav Vestn* 2009; 78 (1): 235-6.

[45] www.origio.com.

[46] Pregl Breznik, B., Kovačič, B., Vlaisavljević, V. Are sperm DNA fragmentation, hyperactivation and hyaluronan-binding ability predictive for fertilization and embryo development in in vitro fertilization and intracytoplasmic sperm injection? *Fertil. Steril.* 2013; 99 (5): 1233-41.

[47] Tournaye, H. How to define limits between IVF and ICSI? 22th annual meeting ESHRE 2006, Prague, 18th - 21st June 2006.

[48] De Mouzon, J., Goosens, V., Bhatacharrya, S., Castilla, J. A., Ferraretti, A. P., Korsak, V., et al. Assisted reproductive technology in Europe, 2006: results generated from European registers by ESHRE. *Hum. Reprod.* 2010; 25: 1851-62.

[49] Saleh, R. A., Agarwal, A., Nelson, D. R., Nada, E. A., El-Tonsy, M. H., Alvarez, J. G., et al. Increased sperm nuclear DNA damage in normozoospermic infertile men: a prospective study. *Fertil. Steril.* 2002; 78: 313–8.

[50] Sakkas, D., Urner, F., Bianchi, P. G., Bizzaro, D., Wagner, I., Jaquenoud, N., et al. Sperm chromatin abnormalities can influence decondensation after intracytoplasmic sperm injection. *Hum. Reprod.* 1996; 11: 837–43.

[51] Hinting, A., Comhaire, F., Vermeulen, L., Dhont, M., Vermeulen, A., Vendakerchkove, D. Value of sperm characteristics and the result of in-

vitro fertilization for predicting the outcome of assisted reproduction. *Int. J. Androl.* 1990; 13: 59-66.

[52] Chen, X., Zhang, W., Luo, Y., Long, X., Sun, X. Predictive value of semen parameters in in vitro fertilization pregnancy outcome. *Andrologia* 2009; 41: 111-7.

[53] Keegan, B. R., Barton, S., Sanchez, X., Berkeley, A. S., Krey, L. C., Grifo, J. Isolated teratozoospermia does not effect in vitro fertilization outcome and is not an indication for intracytoplasmic sperm injection. *Fertil. Steril.* 2007; 88: 1583-8.

[54] Agarwal, A., Allamaneni, S. S. Sperm DNA damage assessment: a test whose time has come. *Fertil. Steril.* 2005; 84: 850-3.

[55] Henkel, R. R., Schill, W. B. Sperm preparation for ART. *Reprod. Biol. Endocrinol.* 2003, 1: 108.

[56] Riffo, M. S., Parraga, M. Study of acrosome reaction and the fertilizing ability of hamster epididymal cauda spermatozoa treated with antibodies against phospholipase A2 and/or lysophosphatidylcholine. *J. Exp. Zool.* 1996; 275 (6): 459-68.

[57] Cocuzza, M., Sikka, S. C., Athayde, K. S., Agarwal, A. Clinical relevance of oxidative stress and sperm chromatin damage in male infertility: An evidence based analysis. *International Braz. J. Urol.* 2007; 33 (5): 603-21.

[58] Sikka, S. C., Rajasekaran, M., Hellstrom, W. J. Role of oxidative stress and antioxidants in male infertility. *J. Androl.* 1995; 16 (6): 464-8.

[59] Johnson, M. H. Coitus and Fertilization. In: Noyes, V., Moore, K., eds. *Essential reproduction.* Avstralija: Blackwell Publishing; 2007: 168-88.

[60] Banks, S., King, S. A., Irvine, D. S., Saunders, P. T. Impact of a mild scrotal heat stress on DNK integrity in murine spermatozoa. *Reproduction* 2005; 129: 505-14.

[61] Bedford, J. M. Effects of elevated temperature on the epididymis and testis: experimental studies. *Advances in experimental medicine and biology* 1991; 286: 19-32.

[62] Bedford, J. M. The status and the state of the human epididymis. *Hum. Reprod.* 1994; 9: 2187-99.

[63] Wolff, H. The biologic significance of white blood cells in semen. *Fertil. Steril.* 1995; 63 (6): 1143-57.

[64] Tremellen, K. Oxidative stress and male infertility-a clinical perspective. *Hum. Reprod. Update* 2008; 43 (3): 243-58.

[65] Armstrong, J. S., Bivalacqua, T. J., Chamulitrat, W., Sikka, S., Hellstrom, W. J. A comparison of the NADHP oxidase in human sperm and white blood cells. *Int. J. Androl.* 2002; 25: 223-9.

[66] Traber, M. G., van der Vliet, A., Reznick, A. Z., Cross, C. E. Tobacco-related disease. Is there a role for antioxidant micronutrient supplementation? *Clin. Chest Med.* 2000; 21 (1): 173-87.

[67] Saleh, R. A., Agarwal, A., Sharma, R. K., Nelson, D. R., Thomas, A. J. Effect of cigarette smoking on levels of seminal oxidative stress in infertile men: a prospective study. *Fertil. Steril.* 2002; 78 (3): 491-9.

[68] Vine, M. F. Smoking and male reproduction: a review. *Int. J. Androl.* 1996; 19 (6): 323-37.

[69] Koch, O. R., Pani, G., Borrello, S., Colavitti, R., Cravero, A., Farrè, S., Galeotti, T. Oxidative stress and antioxidant defences in ethanol-induced cell-injury. *Mol. Aspests Med.* 2004; 25 (1-2): 191-8.

[70] Huszar, G., Vigue, L., Corrales, M. Sperm creatine kinase activity in fertile and infertile oligospermic men. *J. Androl.* 1990; 11: 40-6.

[71] Ergur, A. R., Dokras, A., Giraldo, J. L., Habana, A., Kovanci, E., Huszar, G. Sperm maturity and treatment choice of in vitro fertilization (IVF) or intracytoplasmic sperm injection: diminished sperm HspA2 chaperone levels predict IVF failure. *Fertil. Steril.* 2002; 77: 910-18.

[72] Haaf, F., Sanner, A., Straub, F. Polymers of N-Vinylpyrrolidone: Synthesis, Characterization and Uses. *Polymer Journal* 1985; 17 (1): 143–152.

[73] Hlinka, D., Herman, M., Vesela, J., Hredzak, R., Horvath, S., Pacin, J. A modified method of intracytoplasmic sperm injection without the use of polyvinylpyrrolidone. *Hum. Reprod.* 1998; 13: 1922–7.

[74] Strehler, E., Baccetti, B., Sterzik, K., Capitani, S., Collodel, G., De Santo, M., Gambera, L., Piomboni, P. Detrimental effects of polyvinyl-pyrrolidone on the ultra structure of spermatozoa. *Hum. Reprod.* 1998; 13: 120-3.

[75] Dozortsev, D. P., Ryubouchkin, A., De Sutter, P., Dhont, M. Sperm plasma membrane damage prior to intracytoplasmic sperm injection: a necessary condition for sperm nucleus decondensation. *Hum. Reprod.* 1995;11: 2960–4.

[76] Tsai, M. Y., Huang, F. J., Kung, F. T., Lin, Y. C., Chang, S. Y., Wu, J. F., Chang, H. W. Influence of polyvinylpyrrolidone on the outcome of intracytoplasmic sperm injection. *J. Reprod. Med.* 2000; 45: 115-20.

[77] Mizuno, K., Hoshi, K., Huang, T. Fertilization and embryo development in a mouse ICSI model using human and mouse sperm after immobilization in polyvinylpyrrolidone. *Hum. Reprod.* 2002; 9: 2350-5.

[78] Kato, Y., Nagao, Y. Effect of PVP on sperm capacitation status and embryonic development in cattle. *Theriogenology* 2009; 72: 624-35.

[79] Feichtinger, W., Obruca, A., Brunner, M. Sex chromosomal abnormalities and intracytoplasmic sperm injection. *Lancet* 1995; 346: 1566.

[80] Jean, M., Mirallie, S., Boudineau, M., Tatin, C., Barriere, P. Intracytoplasmic sperm injection with polyvinylpyrrolidone: a potential risk. *Fertil. Steril.* 2001; 76: 419-20.

[81] Suchanek, E., Simunic, V., Juretic, D., Grizelj, V. Follicular fluid contents of hyaluronic acid, follicle-stimulating hormone and steroids relative to the success of in vitro fertilization of human oocytes. *Fertil. Steril.* 1994; 62: 347-52.

[82] Salamonsen, L. A., Shuster, S., Stern, R. Distribution of hyaluronan in human endometrium across the menstrual cycle. Implication for implantation and menstruation. *Cell Tissue Res.* 2001; 306: 335-40.

[83] Campbell, S., Swann, H. R., Aplin, J. D., Seif, M. W., Kimber, S. J., Elstein, M. CD44 is expressed throughout preimplantation human embryo development. *Hum. Reprod.* 1995; 10: 425-30.

[84] Yaegashi, N., Fujita, N., Yajima, A., Nakamura, M. Menstrual cycle dependent expression of CD44 in normal human endometrium. *Hum. Pathol.* 1995; 26: 862-5.

[85] Gardner, D. K., Rodriegoz-Martinez, H., Lane, M. Fetal development After transfer is increased by replacing protein with the glycosaminoglycan hyaluronan for mouse embryo culture and transfer. *Hum. Reprod.* 1999; 14: 2575-80.

[86] Turley, E., Moore, D. Hyaluronate binding protein also bind to fibronectin, laminin and collagen. *Biochem. Biophys. Res. Commun.* 1984; 121: 808-14.

[87] Nakagawa, K., Takahashi, C., Nishi, Y., Jyuen, H., Sugiyama, R., Kuribayashi, Y., Sugiyama, R. Hyaluronan-enriched transfer medium improves outcome in patients with multiple embryo transfer failures. *J. Assist. Reprod. Genet.* 2012; 29 (7): 679-85.

[88] Korosec, S., Virant-Klun, I., Tomazevic, T., Zech, N. H., Meden-Vrtovec, H. Single fresh and frozen-thawed blastocyst transfer using hyaluronan-rich transfer medium. *Reprod. Biomed. Online* 2007;15(6): 701-7.

[89] Loutradi, K. E., Prassas, I., Bili, E., Sanopoulou, T., Bontis, I., Tarlatzis, B. C. Evaluation of a transfer medium containing high concentration of hyaluronan in human in vitro fertilization. *Fertil. Steril.* 2007; 87: 48-52.

[90] Valojerdi, M. R., Karimian, L., Yazdi, P. E., Gilani, M. A., Madani, T., Baghestani, A. R. Efficacy of human embryo transfer medium: a prospective, randomized clinical trial study. *J. assist. Reprod. Genet.* 2006; 23: 207-212.

[91] Check, J. H., Summers-Chase, D., Yuan, W., Swenson, K., Horwath, D., Press, M. "Embryo glue" does not seem to improve chances of subsequent pregnancy in refractory in vitro fertilization cases. *Clin. Exp. Obstet. Gynecol.* 2012; 39(1): 11-2.

In: Hyaluronan
Editor: Vitor H. Pomin

ISBN: 978-1-63117-808-5
© 2014 Nova Science Publishers, Inc.

Chapter 7

HYALURONAN-COATED BONE GRAFT SUBSTITUTES AND THEIR OSTEOCONDUCTIVE PROPERTIES

Stephan Reitinger[1], Robert G. Stigler[2]
and Günter Lepperdinger[1]*
[1]Institute for Biomedical Aging Research,
University of Innsbruck, Australia
[2]Department of Oral and Maxillofacial Surgery,
Medical University Innsbruck, Australia

ABSTRACT

Bone serves as a structural framework and provides protection to various organs, yet also contributes substantially to hemopoiesis and mineral homeostasis. Osseous tissue is densely packed with blood vessels and nerves, which are embedded in an extracellular matrix composed of collagen fibers and crystallized calcium-phosphate salts known as hydroxyapatite.

After a bone fracture, blood typically leaks from the torn ends of vessels and forms a clotted mass known as the fracture hematoma. During healing, fibroblasts, chondroblasts, and osteogenic cells invade the fracture site and start forming a cartilage-and hyaluronan-rich fibrous callus, which is replaced first by spongy bone trabeculae and

---

* E-mail address: stephan.reitinger@uibk.ac.at.

subsequently is remodeled to compact bone. Depending on the site and size of the bone defect and on the health status and age of the patient, complete healing can require weeks to several months.

In this regard, treatments with β-tricalcium phosphate-based bone graft substitutes have been reported to support the overall regeneration progress. Further along these lines, hyaluronan-coated biocompatible bone replacement material provides a more favorable environment for invading cells and was shown to serve as a delivery vehicle for autologous cells. In this review, the osteoconductive role of hyaluronan and its medical application in bone graft substitutes in the context of fracture healing, bone fragment fixation surgery and implant osseointegration will be discussed.

INTRODUCTION

The high regenerative capacity of bone is well acknowledged and is reflected in the fact that most skeletal fractures will heal without broad medical intervention. Bone tissue is subject to a massive turnover rate allowing a substantial amount of the human skeleton to be renewed every year. In accordance with its high metabolic activity, fractures heal without forming fibrous scars. This type of mending is therefore often compared to those processes taking place during bone development.

Based on the solidity and rigidity of fixation of fractured bone fragments, a direct cortical (primary) or indirect (secondary) healing pattern is observed [1]. Primary osteonal reconstruction, which does not show intermediate callus formation, requires the correct anatomical reduction of the fracture ends as well as a highly stable fixation. Both requirements are typical for gap or contact healing. The major goal after internal fixation surgery is direct remodeling of the structurally and functionally competent lamellar bone architecture in order to directly reestablish the cortical continuity. However, direct remodeling is rarely observed in the natural process of fracture healing [2]. More prevalent is indirect fracture healing, a process characterized by consecutive yet overlapping phases [3]. These are characterized by initial inflammation, which is followed by repair. Newly formed bone is eventually taken into the same type of remodeling as previously established bone. Immediately following the trauma and as a direct consequence of damaged blood vessels, a coagulated hematoma is generated, which contains various cell types, some of which secrete inflammatory cytokines. This proinflammatory environment attracts further immune and bone marrow cells and stimulates the production of an extracellular matrix (ECM), which promotes neovascularization as well as the proliferation of mesenchymal cells, which commit themselves to differentiate into chondrocytes and osteoprogenitors.

In the context of effective fracture healing, the presence of a scaffold, together with respective ECM, appears to be imperative in order to present osteoinductive stimuli [4]. A well-studied, major component of the ECM not only at sites of cartilaginous and osseous remodeling [5], but in all vertebrate tissues and fluids, is the glycosaminoglycan hyaluronan (HA) [6]. This unbranched high-molecular mass biopolymer directly interacts with cells via the cell surface receptors CD44 [7] or HA-mediated motility receptor (HMMR) [8]. HA-accumulations hold large volumes of water, thereby causing local swelling and clearing space at the fracture site. This process allows bone marrow cells to invade, divide, and differentiate [9]. The physiochemical properties of HA, which first and foremost promote a high extent of hydration, greatly support cellular processes during blood vessel formation and neovascularization of injured areas [10, 11]. Most importantly to note here, HA is actually being synthesized in cartilaginous and osseous tissue [5], which are primary components forming soft and hard callus during fracture healing. HA constitutes also a major element of the shared niche of mesenchymal stem cells (MSC) and hematopoietic stem cells (HSC) [12, 13].

Because of its important role in the context of bone healing, the use of HA-loaded bone graft materials in osseous surgery is being broadly investigated by basic researchers and clinicians and is considered to be of high therapeutic relevance, in particular, in an osteoporotic situation, as well as in non-unions or large critical-size defects.

OSTEOCONDUCTIVE BONE GRAFT MATERIALS

Considerable efforts are being made to engineer and develop bone graft substitute material not only to repair or regenerate bone defects, but also to compensate for bone loss through disease or surgery after cancer or injury. For smaller defects, as have been commonly observed in maxillofacial surgery, bone replacement material can be used. For larger defects autologous bone is needed, which is either devoid of vascular structures or a transplant carrying microvasculature suitable for later surgical anastomosis. Autologous bone grafts, i. e. osseous specimens from the patient's own body, are harvested from non-essential, non-load-bearing bone sites, providing material for augmentation or replacement. However, it involves an additional surgical intervention and thus may increase the risk for post-operative complications at another site of the body [14]. For defects where large amounts of graft volume are needed, demineralized bone matrix (DBM) is mixed with available autologous osseous tissue before employing it as a bone graft extender. DBM is generated by processing (demineralizing) allograft bone specimen derived from cadavers or bone banks. It still contains endogenous proteins, which may impart the benefits of osteoinductivity in the transplant environment [15]. Risks and complications with allografts include bacterial and viral infections and non-unions [16].

Over the last centuries surgeons have not only used auto-, allo- or even xenografts to repair bone defects but also applied a variety of artificial materials such as metal and plastics [17]. Substitution of DBM or corticocancellous grafts by application of synthetic materials, often also referred to as "artificial bone", is a previously established research field with broad clinical implications for osseous surgery [17]. An ideal bone graft material should not only be biocompatible and integrate into the existing bony tissue, but needs to provide osteoinductive properties in order to support new bone formation [16, 18]. To name only a few very new developments, silicon, bioactive glasses and glass ionomers, have all been shown to exhibit osteoinductive properties. More traditionally implemented are calcium phosphates such as β-tricalcium phosphate (βTCP) or synthetic hydroxyapatite [16].

These bone substitutes generally serve as fillers. However, since the materials themselves show only little biological activity, attemps to increase bioactivity were undertaken by impregnating them with cell-stimulating substances, in order to engineer more bioactive bone graft materials. In this capacity, the biochemical component HA has been well studied [19]. This natural biopolymer has gained wide clinical approval and is frequently used in hydrogels, in particular, in its use as a topical wound medication [20].

HA MEDIATES OSTEOGENESIS

In bone, HA's unique hydrodynamic properties influence MSC, HSC as well as osteogenic cells. Therefore its distribution has been suggested to be crucial for proper tissue functions [21]. The polysaccharide is abundant in the shared MSC/HSC bone marrow niche [12]. There it is considered to be a key component governing stem cell behavior [13, 22] and HSC homing via the HA adhesion receptor CD44 [23]. In the presence of HA cultivated MSC show a slow-cycling mode exhibiting a prolonged G_1-phase and a suppressed S-phase entry [24], features typical for dormant stem cells and crucial for maintaining stemness. In this context, in vitro expanded bone marrow MSC have been found to express hyaluronan synthase isozymes HAS1, HAS2, and HAS3 and they efficiently form CD44-mediated pericellular HA-coats [25]. In HAS2-deficient mice, impaired growth plate formation and skeletal patterning during bone development was observed [26]. Furthermore, the authors claimed that the HAS2-mediated production of HA is required for cellular organization within osseous tissue, as well as to enhance chondrocyte maturation. In line with this finding, HAS2 overexpression in chick limb development resulted in malformed limbs and missing skeletal elements [27]. The study showed that timely downregulation of HA is crucial for the proper formation of pre-cartilage condensation and subsequent chondrogenesis. Detailed analysis of experiments in a rat calvaria model identified HAS2 as being a leukemia inhibitory factor (LIF)-induced gene [28]. The cytokine LIF acts via the heterodimeric receptor gp130/gp190 and has been shown to decrease mineral deposits during osteogenesis in a dose-dependent manner, and as a consequence thereof interferes with osteoblast differentiation through stimulating HAS2 activity and concomitantly increasing HA levels [29, 30]. In addition to the spatiotemporal activation of HA synthesis, it has been assumed that the polymeric length of newly produced HA is of importance [31]. High molecular mass HA was found to act as a cell-encircling barrier that blocks cell migration, which can be restored by enzymatic HA degradation through hyaluronidase treatment [32, 33]. Short HA fragments have been reported to increase blood flow to sites of vascular injury and to stimulate angiogenesis [34]. Conclusively, a concerted interplay between HA production and degradation is needed not only during development but also to effectively restore damaged osseous architecture and function [35].

APPLICATION OF HA IN OSSEOUS TISSUE REGENERATION

In orthopaedics, sodium hyaluronate is often injected intra-articularly to act as a lubricant and/or as pain reliever. The efficacy of this treatment is presently still controversial [36, 37]. Based on an assumption anticipating the decisive osteoinductivity of HA, cavities in injured bone marrow in rats were filled with high molecular weight HA. The therapy resulted in enhanced trabecular bone formation [38]. At this time it was interpreted that HA potentially serves as a buffer for growth factors, which effectively stimulate bone progenitors to proliferate and differentiate. Yet, in particular, low molecular weight HA fragments have been shown to effectively stimulate neovascularization [39]. The combination of DMB and HA did indeed increase total vessel number in a rat tibial bone marrow ablation model [11]. This concept was also tested using special matrices that were based on chemically modified, crosslinked HA, such as HYAFF11. They were first employed to test binding of mesenchymal cells, promoting faster and supposedly also a more complete regeneration of the graft. The HA matrices were found to perform better than porous calcium phosphate ceramics, composed of 60% hydroxyapatite and 40% TCP. Yet, in these studies, the combination of both has not been tested [40]. Therefore, HYAFF11 was formulated together with α-TCP and hydroxyapatite, thereby lending great support to the view that this type of composite exhibits superior biomechanical properties, suitable for the use as a filler [41]. In order to validate this concept in a clinical trial, DMB with an additional bone substitute material of non-human origin such as Bio-Oss, which at that time was commonly used in trauma surgery, was coated with HA. The composite material thus gained a putty-like consistency, which notably was also much easier to handle during its application in sinus lift augmentation. Grafting HA-coated DMB resulted in remarkable retention of the implant material with the operation site and thus yielded more efficient new bone formation [42]. HA in combination with combined βTCP granules has been implanted in rabbit femoral condyles, βTCP blocks were used as controls. With respect to osteoconductive properties, the injectable paste performed similarly well [43]. This biomaterial formulation has been further refined by adding methyl cellulose in order to provide an osteoinductive material, which simultaneously exhibits immune modulatory properties and stimulates vascularization. Working along this line, a coherent sticky material was composed, which, when subjected subcutaneously in a mouse model, refrained from merging with the surrounding tissue, while concomitantly inducing enhanced vascularization with the immediate surroundings [44].

Intentional biofunctionalization is used to specifically enhance the beneficial properties of a biomaterial. In the context of osseous defect healing, bonemorphogenetic proteins such as BMP2 or BMP7 are often applied and have also been clinically tested. In this respect, HA matrices with variant release kinetics for incorporated growth factors such as BMP2 and the proangiogenic factor vascular endothelial growth factor (VEGF) were tested in a rat calvaria defect model. These experiments showed a great enhancement in healing effects, in particular, when co-delivering VEGF [45]. These developments have been taken further to improve osteochondral defect healing. Damaged articular cartilage has a limited capacity to heal. Therefore, biphasic composites were developed, in which the osseous phase consists of βTCP/hydroxyapatite and for the chondral phase HA is combined with atelocollagen. The fabricated biphasic scaffolds were tested in osteochondral defects in the knee joint of miniature pigs. The study showed that diligently engineered composite material may provide a basis to implanted chondrocytes, in order to form load bearing and functional tissue by stabilization the underlying bone through concomitantly efficient bone healing [46].

SUMMARY AND OUTLOOK

The role of HA as a mediator of basic cellular properties and physiologic characteristics, which guide and control proliferation and differentiation has been well-studied. Due to its ubiquitous presence in the human body, it is considered key in somatic maintenance, repair, regeneration and remodeling of organs and tissues. Along these lines, HA was also found to regulate the basic mechanisms of stem cells within bone tissue, in particular within the bone marrow. There, it appears to be functionally and structurally involved in maintaining proper osseous tissue function as well as playing a role in fostering bone repair and regeneration. Therefore the formulation of HA into a viscous solution or into a hydrogel has been implemented in novel therapeutic strategies and has indeed been shown to exhibit greatly enhancing effects in the process of bone healing.

Presently, research and development, which is to bring forth innovative therapeutic solutions is promoting the production of complex multi-phasic materials. Primarily these materials are based on calcium phosphate as a scaffold, however they are being biofunctionalized in a way enabling them to release cell-stimulating cytokines and growth factors. Above all, for these synthetic matrices to be therapeutically relevant, they must release those stimulating factors in a predicable manner. Taking the performance of HA into account, be it the unrefined high molecular weight polysaccharide or be it special formulations such as reticulated or well-defined chemical derivatives of the polymer, HA appears to be well suited for clinical applications promoting the osteoconductive processes and may eventually play a role in achieving complete bone healing.

ACKNOWLEDGMENTS

We are grateful to Kai Kosog for critically reading this manuscript. Funding by the European Commission is gratefully acknowledged: SR is supported by the Marie Curie International Reintegration Grant Project HyalStemAge (FP7-PEOPLE-2010-RG-277085) and GL by the EC FP7 integrated project VascuBone (FP7-HEALTH-2009-242175).

REFERENCES

[1]   P. Kolar, K. Schmidt-Bleek, H. Schell, T. Gaber, D. Toben, G. Schmidmaier, C. Perka, F. Buttgereit, and G. N. Duda, *Tissue Engineering. Part B, Reviews.* 16, 427 (2010).

[2]   R. Marsell and T. A. Einhorn, *Injury.* 42, 551 (2011).

[3]   L. Claes, S. Recknagel, and A. Ignatius, *Nature Reviews. Rheumatology.* 8, 133 (2012).

[4]   D. E. Komatsu and S. J. Warden, *Journal of Cellular Biochemistry.* 109, 302 (2010).

[5]   E. R. Bastow, S. Byers, S. B. Golub, C. E. Clarkin, A. A. Pitsillides, and A. J. Fosang, *Cellular and Molecular Life Sciences : CMLS.* 65, 395 (2008).

[6]   H. G. Garg and C. A. Hales, *Chemistry and Biology of Hyaluronan*, Elsevier Ltd, Oxford (2004).

[7]   C. Underhill, *Journal of Cell Science.* 103 (Pt 2), 293 (1992).

[8]   E. A. Turley, L. Austen, K. Vandeligt, and C. Clary, *The Journal of Cell Biology.* 112, 1041 (1991).

[9]   A. S. Hoffman, *Advanced Drug Delivery Reviews.* 54, 3 (2002).

[10]   F. Gao, C. X. Yang, W. Mo, Y. W. Liu, and Y. Q. He, *Clinical and investigative medicine. Medecine Clinique et Experimentale.* 31, E106 (2008).

[11]   A. L. Raines, M. Sunwoo, A. A. Gertzman, K. Thacker, R. E. Guldberg, Z. Schwartz, and B. D. Boyan, *Journal of Biomedical Materials Research. Part A.* 96, 575 (2011).

[12]   S. Méndez-Ferrer, T. V. Michurina, F. Ferraro, A. R. Mazloom, B. D. Macarthur, S. A. Lira, D. T. Scadden, A. Ma'ayan, G. N. Enikolopov, and P. S. Frenette, *Nature.* 466, 829 (2010).

[13] M. A. Solis, Y. H. Chen, T. Y. Wong, V. Z. Bittencourt, Y. C. Lin, and L. L. Huang, *Biochemistry Research International.* 2012, 346972 (2012).

[14] S. T. Becker, P. H. Warnke, E. Behrens, and J. Wiltfang, *Journal of Oral and Maxillofacial Surgery.* 69, 48 (2011).

[15] G. Zimmermann and A. Moghaddam, *Injury.* 42 Suppl 2, S16 (2011).

[16] W. R. Moore, S. E. Graves, and G. I. Bain, *ANZ Journal of Surgery.* 71, 354 (2001).

[17] A. Sanan and S. J. Haines, *Neurosurgery.* 40, 588 (1997).

[18] G. M. Calori, E. Mazza, M. Colombo, and C. Ripamonti, *Injury.* 42 Suppl 2, S56 (2011).

[19] L. Sorokin, *Nature reviews. Immunology.* 10, 712 (2010).

[20] J. A. Burdick and G. D. Prestwich, *Advanced Materials.* 23, H41 (2011).

[21] G. Sundström, I. M. Dahl, M. Hultdin, B. Lundström, A. Wahlin, and A. Engström-Laurent, *Medical Oncology.* 22, 71 (2005).

[22] V. Goncharova, N. Serobyan, S. Iizuka, I. Schraufstatter, A. de Ridder, T. Povaliy, V. Wacker, N. Itano, K. Kimata, I. A. Orlovskaja, Y. Yamaguchi, and S. Khaldoyanidi, *The Journal of Biological Chemistry.* 287, 25419 (2012).

[23] A. Avigdor, P. Goichberg, S. Shivtiel, A. Dar, A. Peled, S. Samira, O. Kollet, R. Hershkoviz, R. Alon, I. Hardan, H. Ben-Hur, D. Naor, A. Nagler, and T. Lapidot, *Blood.* 103, 2981 (2004).

[24] C. M. Liu, C. H. Yu, C. H. Chang, C. C. Hsu, and L. L. Huang, *Cell and Tissue Research.* 334, 435 (2008).

[25] C. Qu, K. Rilla, R. Tammi, M. Tammi, H. Kroger, and M. J. Lammi, *The International Journal of Biochemistry & Cell Biology.* 48C, 45 (2014).

[26] K. Matsumoto, Y. Li, C. Jakuba, Y. Sugiyama, T. Sayo, M. Okuno, C. N. Dealy, B. P. Toole, J. Takeda, Y. Yamaguchi, and R. A. Kosher, *Development.* 136, 2825 (2009).

[27] Y. Li, B. P. Toole, C. N. Dealy, and R. A. Kosher, *Developmental Biology.* 305, 411 (2007).

[28] D. Falconi and J. E. Aubin, *Journal of Bone and Mineral Research.* 22, 1289 (2007).

[29] S. Bohic, P. Pilet, and D. Heymann, *Biochemical and Biophysical Research Communications.* 253, 506 (1998).

[30] L. Malaval, A. K. Gupta, and J. E. Aubin, *Endocrinology.* 136, 1411 (1995).

[31] R. Stern, A. A. Asari, and K. N. Sugahara, *European Journal of Cell Biology.* 85, 699 (2006).

[32] C. Tempel, A. Gilead, and M. Neeman, *Biology of reproduction.* 63, 134 (2000).

[33] B. P. Toole, *Seminars in Cell & Developmental Biology.* 12, 79 (2001).

[34] V. C. Lees, T. P. Fan, and D. C. West, *Laboratory Investigation.* 73, 259 (1995).

[35] D. D. Allison and K. J. Grande-Allen, *Tissue engineering.* 12, 2131 (2006).

[36] J. Arrich, F. Piribauer, P. Mad, D. Schmid, K. Klaushofer, and M. Mullner, *CMAJ : Canadian Medical Association Journal.* 172, 1039 (2005).

[37] E. A. Balazs, *Surgical Technology International.* 12, 278 (2004).

[38] T. Sasaki and C. Watanabe, *Bone.* 16, 9 (1995).

[39] D. C. West, I. N. Hampson, F. Arnold, and S. Kumar, *Science.* 228, 1324 (1985).

[40] L. A. Solchaga, J. E. Dennis, V. M. Goldberg, and A. I. Caplan, *Journal of Orthopaedic Research.* 17, 205 (1999).

[41] V. Sanginario, M. P. Ginebra, K. E. Tanner, J. A. Planell, and L. Ambrosio, *Journal of materials science. Materials in Medicine.* 17, 447 (2006).

[42] Z. Schwartz, M. Goldstein, E. Raviv, A. Hirsch, D. M. Ranly, and B. D. Boyan, *Clinical oral Implants Research.* 18, 204 (2007).

[43] M. Chazono, T. Tanaka, H. Komaki, and K. Fujii, *Journal of Biomedical Materials Research. Part A.* 70, 542 (2004).

[44] S. Ghanaati, M. Barbeck, U. Hilbig, C. Hoffmann, R. E. Unger, R. A. Sader, F. Peters, and C. J. Kirkpatrick, *Acta Biomaterialia.* 7, 4018 (2011).

[45] J. Patterson, R. Siew, S. W. Herring, A. S. Lin, R. Guldberg, and P. S. Stayton, *Biomaterials.* 31, 6772 (2010).

[46] G. I. Im, J. H. Ahn, S. Y. Kim, B. S. Choi, and S. W. Lee, *Tissue Engineering. Part A.* 16, 1189 (2010).

EDITOR'S CONTACT INFORMATION

Dr. Vitor H. Pomin
Professor
Biochemistry and Chemical Biology,
Institute of Medical Biochemistry,
Federal University of Rio de Janeiro,
Rio de Janeiro, Brazil
Tel.: +55-21-2562-2939
pominvh@bioqmed. ufrj .br

INDEX

F

G

N

T